ISO 27001 Controls

A guide to implementing and auditing

Second edition

ISO 27001 Controls

A guide to implementing and auditing

Second edition

BRIDGET KENYON

IT Governance Publishing

IT Governance Publishing Ltd
Unit 3, Clive Court
Bartholomew's Walk
Cambridgeshire Business Park
Ely, Cambridgeshire
CB7 4EA
United Kingdom
www.itgovernancepublishing.co.uk

© Bridget Kenyon 2019, 2024.

The author has asserted the rights of the author under the Copyright, Designs and Patents Act, 1988, to be identified as the author of this work.

Formerly published as *Guide to the Implementation and Auditing of ISMS Controls based on ISO/IEC 27001* by BSI.

First published in the United Kingdom in 2019 by IT Governance Publishing.

ISBN 978-1-78778-144-3

This edition published in the United Kingdom in 2024 by IT Governance Publishing.

ISBN 978-1-78778-430-7

Cover image originally sourced from Shutterstock®.

ABOUT THE AUTHOR

Bridget Kenyon is the Chief Information Security Officer for SSCL. She is responsible for managing strategy and information security activities for the whole organisation, including internal and customer-facing elements. In parallel to her main role, Bridget has been on the ISO editing team for ISMS standards since 2006; she has served as Lead Editor for ISO/IEC 27001:2022 and ISO/IEC 27014:2020.

Joining DERA in 2000 to work on network vulnerabilities, Bridget discovered her passion for information and cyber security. Following this, she has been a Qualified Security Assessor against PCI DSS, Head of Information Security for UCL, and has held operational and consultancy roles in both industry and academia. She is a member of the UK Advisory Council for ISC2, and a Fellow of the Chartered Institute of Information Security.

Bridget's interests lie in finding the edges of security which you can peel up, and the human aspects of system vulnerability. She's the sort of person who will always have a foot in both the technical and strategy camps. She enjoys helping people to find solutions to thorny problems, and strongly believes that cyber and information security are fundamental to resilient business operations, not "nice to haves".

DISCLAIMER

A document such as this is provided with the best of intentions. It reflects publicly available common best practice, which is derived from a consensus among international experts with a wide variety of skills, knowledge and experience in the subject. This guide makes no claim to be exhaustive or definitive, and users may need to seek further guidance in implementing the requirements of ISO/IEC 27001 or the use of the guidance found in ISO/IEC 27002. Furthermore, there will always be other aspects where additional guidance is required relevant to the organisational, operational, legal and environmental context of the business, including specific threats, controls, regulatory compliance, governance and good practice.

The author of this guide cannot be held liable by organisations, users or third parties for the execution or implementation of this information. It has been assumed in drafting the information and advice given in this guide that the execution of this information by organisations and users is entrusted to appropriately qualified and experienced people.

Unless stated otherwise, all quotations are from ISO/IEC 27001:2022.

viii

CONTENTS

Contents

Contents

Contents

Contents

Contents

FOREWORD

Information is one of your organisation's most valuable assets. The objectives of information security are to protect the confidentiality, integrity and availability of information. These basic elements of information security help ensure that an organisation can protect against:

- sensitive or confidential information being given away, leaked or otherwise exposed, either accidentally or deliberately;
- personally identifiable information being compromised;
- critical information being accidentally or intentionally modified without the organisation's knowledge;
- important business information being lost without a trace or hope of recovery; and
- important business information being unavailable when needed.

It should be the responsibility of all managers, information system owners or custodians, and users in general to ensure that their information is properly managed and protected from the variety of risks and threats faced by every organisation. The two standards ISO/IEC 27001:2022, *Information security, cybersecurity and privacy protection – Information security management systems – Requirements*, and ISO/IEC 27002:2022, *Information security, cybersecurity and privacy protection – Information security controls*, together provide a basis for organisations to develop an effective information security management framework for managing and protecting their important business assets, while minimising their risks, maximising investment and business opportunities, and ensuring their information systems continue to be available and operational.

ISO/IEC 27001 is a requirements standard that can be used for accredited third-party information security management system (ISMS) certification. Organisations going through the accredited certification route have their ISMS audited by an accredited certification body. This ensures that they have appropriate management processes and systems in place, and that these conform to the requirements specified in ISO/IEC 27001.

ISO/IEC 27002, a guidance document, provides a comprehensive set of best-practice controls for information security and implementation guidance. Organisations can adopt these controls as part of the risk treatment process specified in ISO/IEC 27001 to manage the risks they face to their information assets.

This guide is designed to help you establish, implement and maintain your ISMS to help you prepare for ISMS certification. This is the fifth edition of this book, and it has been updated to reflect the publication of the latest versions of ISO/IEC 27001 and 27002 in 2022.

NOTE: A book such as this is provided with the best of intentions. It reflects publicly available common best practice, which is derived from a consensus among international experts with a wide variety of skills, knowledge and experience in the subject. This guide makes no claim to be exhaustive or definitive, and users may need to seek further guidance in implementing the requirements of ISO/IEC 27001 or the use of the guidance in *ISO/IEC 27002. Furthermore, there will always be other aspects where additional guidance is required relevant to the organisational, operational, legal and environmental context of the business, including specific threats, controls, regulatory compliance, governance and good practice. The author of this guide cannot be held liable by organisations, users or third parties for the execution or implementation of this information. The author has assumed that the execution of this information by organisations and users is entrusted to appropriately qualified and experienced people.*

CHAPTER 1: SCOPE OF THIS GUIDE

This guide provides instructions on implementing ISMS control requirements and on auditing existing control implementations to help organisations prepare for certification in accordance with ISO/IEC 27001.

The guide covers the ISMS control requirements that should be addressed by organisations considering certification. Chapters 5–8 discuss each of the controls in Annex A of ISO/IEC 27001 from two different viewpoints.

- **Implementation guidance** – what needs to be considered to fulfil the control requirements when implementing the controls from Annex A of ISO/IEC 27001. This guidance is aligned with ISO/IEC 27002, which gives advice on implementing the controls.

- **Auditing guidance** – what should be checked, and how, when examining the implementation of ISO/IEC 27001 controls to ensure that the implementation covers the ISMS control requirements.

It is important to emphasise that this guide does not cover the implementation or auditing of the ISMS process requirements (the main body of ISO/IEC 27001). This is discussed in more detail in Chapter 3.

CHAPTER 2: FIELD OF APPLICATION

2.1 Usage

This guide is intended to be used by those involved in:

- designing, implementing and/or maintaining an ISMS;
- preparing for ISMS audits and assessments;
- carrying out internal ISMS audits and assessments[1]; and
- carrying out ISMS audits and assessments of other organisations.

The guide refers to the following standards:

- **ISO/IEC 27001** – the requirements specification for an ISMS. This International Standard is used as the basis for accredited certification.
- **ISO/IEC 27002** – a reference for selecting controls as part of the implementation of an ISMS, and a guidance document for organisations implementing commonly accepted security controls.

The guide will be updated following any changes to these standards. Organisations should therefore ensure that the correct version is being used for compliance checks related to pre-certification, certification and post-certification purposes.

---

[1] Auditors deployed by the organisation to carry out an internal ISMS audit, auditors from certification bodies, and assessors from accreditation bodies engaged in assessing certification bodies.

2.2 Compliance

To claim compliance with the requirements of ISO/IEC 27001, the organisation needs to demonstrate that it has all the processes in place and provides appropriate objective evidence to support such claims. Any exclusion of controls found to be necessary to satisfy the risk acceptance criteria needs to be justified. The organisation should also provide evidence that the associated risks have been knowingly and objectively accepted by those in management who have the executive responsibility and are accountable for making such decisions.

Excluding any of the requirements specified in Clauses 4–10 of ISO/IEC 27001 is not acceptable.[2]

The implementation of ISMS processes results in the organisation deploying a system of controls based on a risk management approach to manage its risks. The organisation should have implemented an effective system of management controls and processes as part of its ISMS, and should be able to demonstrate this by providing evidence to the ISMS auditor (whether it be a first-, second- or third-party audit).

This guide can be used by those who might not have an immediate need for an audit but require a specification for establishing and implementing an ISMS based on industry-accepted best-practice processes. However, claiming compliance with ISO/IEC 27001 requires the organisation to have at least an internal ISMS audit in place, whether or not

---

[2] See ISO/IEC 27001, 1.

it goes for a third-party audit at a later stage. The organisation may not have a business case for a third-party audit, but to be compliant with ISO/IEC 27001, an internal ISMS audit process is mandatory. This guide can, of course, also be used by those preparing for a second-party or third-party audit.

CHAPTER 3: MEETING ISO/IEC 27001 REQUIREMENTS

ISO/IEC 27001 has two main parts.

- The requirements for processes in an ISMS, which are described in Clauses 4–10 (the main body of the text).
- A list of ISMS controls, which is given in Annex A. These controls are described in more detail in ISO/IEC 27002.

The ISMS process requirements address how an organisation should establish and maintain its ISMS. An organisation that wants to achieve ISO/IEC 27001 certification needs to comply with all of these requirements – exclusions are not acceptable.

The ISMS controls listed in Annex A of ISO/IEC 27001 are not mandatory. They are expected to be used as an aide-memoire to help the organisation identify where it might have missed a risk or relevant security control in its risk assessment and the creation of its risk treatment plan. This is stated in ISO/IEC 27001 as follows:

"The organization shall [...] produce a Statement of Applicability that contains the necessary controls [... and] justification for their inclusion; whether the necessary controls are implemented or not; and the justification for excluding any of the Annex A controls."

CHAPTER 4: USING CONTROL ATTRIBUTES

Each control in ISO/IEC 27002 contains a table of metadata pertaining to that control, to enable the categorisation and analysis of control sets; these data are referred to as attributes. ISO/IEC 27002 provides a summary of controls by attribute in Annex A, and describes how to create additional organisation-specific attributes.

The use of attributes is (obviously) optional, and a subset of attributes can be referenced to suit the needs of the organisation.

Attributes can be used in a variety of ways; a few are listed below.

- **Identification of imbalanced controls:** if there are many controls in place that are detective or reactive, but not many that are preventive, this might indicate that the organisation can put more focus on prevention, thus reducing the frequency of incidents.

- **Identification of over-protection:** where an organisation has experienced an incident, there can be a tendency to over-implement controls focusing on the specific incident that occurred. While risk assessment should be capable of highlighting missing or inadequate protections, it will not necessarily identify over-corrected areas. A review of controls by attribute (e.g. the "Operational capabilities" dimension) provides a chance to see them in a different light, and identify where resource can be reallocated without increasing risk.

- **Selection of controls:** where a risk assessment methodology using 'likelihood' and 'impact' (or equivalent) is in use, it can be cross-referenced with the control attributes to select a control that is suitable for the desired effect on risk. For example, if a risk is intended to be controlled by reducing its impact, then a control with the attribute "preventive" would be appropriate.

- **Control review:** where risks and controls are already defined, the attributes can be reviewed to ensure that the required effect on risk is consistent with the control attribute; this will help identify where controls may have been mis-selected in the past.

- **Relationship to other control frameworks:** some control attributes relate to frameworks outside the ISO space, and can hence be used to correlate between the different information and cyber security frameworks in use by the organisation.

CHAPTER 5: ORGANIZATIONAL CONTROLS (ISO/IEC 27001, A.5)

In this section, each of the control objectives and control requirements in Annex A of ISO/IEC 27001 are discussed from both implementation and auditing viewpoints, taking into account the implementation advice given for each control in ISO/IEC 27002.

Readers are encouraged to review both the implementing and auditing sections to understand what is required and how it might be tested.

5.1 Policies for information security (ISO/IEC 27001, A.5.1)

> *"Information security policy and topic-specific policies shall be defined, approved by management, published, communicated to and acknowledged by relevant personnel and relevant interested parties, and reviewed at planned intervals and if significant changes occur."*

Implementation guidance

Guidance on what an information security policy should contain is in ISO/IEC 27002, 5.1.

Organisational policies should be simple and to the point. As it is not appropriate to combine every level of policy into one document, the top-level information security policy should link to more detailed topic-specific policies. Indeed, the top-level policy should normally be just one page long. It might also be part of a more general policy document. The top-level

information security policy should be distributed and
communicated to all staff, and to all relevant external parties,
including others regularly working on the organisation's
premises. The information security policies should be
subject to version control and form part of the ISMS
documentation.

The topic-specific policies should be available to appropriate
staff as needed, dependent on their job function and the
associated security requirements, and classified accordingly.
Information security policies should also be made available
to anyone with appropriate authorisation on request, and they
should be protected from tampering and unintentional
damage. When an information security policy is distributed
outside the organisation, it should be redacted, with any
sensitive information that might be in it removed before such
distribution.

The top-level information security policy and several, or all,
of the lower-level policies could be delivered to staff within
a security policy manual.

Policy review forms an important part of the continual
maintenance, review and updating of the ISMS, which is also
addressed in Clause 9 of ISO/IEC 27001: 'Performance
evaluation'.

This maintenance process should detect all changes that
affect the ISMS, and update the information security policy
and topic-specific policies if appropriate to keep them
current and to ensure that they accurately reflect how the
organisation manages its risks.

Scheduled periodic reviews and defined review procedures
are also essential to ensure that any changes that have gone

undetected thus far are detected and brought into the standard document control process. Appointing an owner for the information security policy, with responsibility for its review, helps ensure that a periodic review takes place.

Staff should also be made aware of policy changes that may affect their role.

Auditing guidance

The top-level information security policy does not need to be extensive but should clearly state senior management's commitment to information security, be under change and version control, and be signed by the appropriate senior manager. The policy should at least address the following topics.

- A comprehensible definition of information security, its overall scope and objectives.
- The reasons information security is important to the organisation.
- A statement of top management's support for information security.
- A summary of the practical framework for risk assessment, risk management, and selecting control objectives and controls.
- A summary of the security policies, principles, standards and compliance requirements.
- A definition of all relevant information security responsibilities (see also 5.2).

- Reference to supporting documentation, e.g. more detailed policies.
- How non-compliances and exceptions will be handled.

The auditor should confirm that the policy is readily accessible to all employees and all relevant external parties, and that it is communicated to all relevant persons, checking that they are aware of its existence and understand its contents. The policy may be a standalone statement or part of more extensive documentation (e.g. a security policy manual) that defines how the information security policy is implemented in the organisation. In general, most, if not all, employees covered by the ISMS's scope will have some responsibility for information security, and auditors should review any declarations to the contrary with care.

The auditor should also confirm that the policy has an owner who is responsible for its maintenance (see also 5.2), and that it is updated appropriately following any changes affecting the information security requirements of the organisation, such as changes in the original risk assessment.

Topic-specific policies that underpin the top-level policy should be clearly linked to the needs of their target group(s), and cover all subjects and concepts that are necessary to provide a foundation for other security controls.

Auditors should confirm that the organisation has appointed an owner for its information security policy with responsibility for its review, or, if this has not taken place, ensured that other clear responsibilities are in place for the review.

Auditors should also confirm that the organisation has developed procedures to react to any incidents, new vulnerabilities or threats, changes in technology, or anything else that is related to the ISMS and might make a review of the policy necessary.

There should also be scheduled periodic reviews to ensure that the policy remains appropriate and is cost-effective to implement in relation to the protection achieved. The auditor should confirm that the schedule for such reviews is appropriate for the overall risk context of the organisation.

Finally, auditors should check the organisation's plans for distributing updated policies and verify that all employees are made aware of the changes.

5.2 Information security roles and responsibilities (ISO/IEC 27001, A.5.2)

"Information security roles and responsibilities shall be defined and allocated according to the organization needs."

Implementation guidance

The organisation will be vulnerable if employees, contractors and third parties are not aware of what rules they must comply with, and what behaviour is expected of them. All employees, contractors and third-party users should have their roles and responsibilities relating to information security defined and clearly communicated to them. These roles and responsibilities should also be consistent with the organisation's security policies.

Responsibility for the protection of individual assets and for carrying out specific security processes should be clearly defined and documented in accordance with information security policies (see 5.9). This is not a trivial task, and should encompass every employee. It is fundamental that management and staff should be told what is expected of them, especially where information security is not likely to be their first interest. In general, all staff should have a basic responsibility for security noted in their job description (see also 6.2), and they should understand their security responsibilities to be an integral part of their job function. Those employees, contractors and third-party users with substantial and complex security responsibilities should have these documented, for example in their job description or in the terms and conditions of employment (see 6.2). This document could be signed by the employee, contractor and third-party user, and their manager, to indicate acceptance and understanding. All employees, contractors and third-party users should be given a personal copy.

When information security responsibilities change, those affected should be made aware of the change in a timely fashion and provided with appropriate training to ensure they can fulfil their changed responsibilities (see 6.3).

Information security responsibilities may also be delegated to others during work processes. In such cases, it is important that the persons delegating their responsibilities are aware that they remain accountable, and that it is part of their role to determine that any delegated tasks and responsibilities have been performed correctly. This also applies when an organisation outsources activities to another entity (see 5.21). A member of the organisation should be responsible for managing the relationship and ensuring that the third party carries out their part of the activities that support the ISMS.

Auditing guidance

Auditors should confirm that somebody with overall responsibility for information security has been appointed (e.g. a Head of Information Security, or Chief Information Security Officer). This role may be separate from the role that has overall ownership of information risk within the organisation.

The information security policy, and/or the risk treatment plan, is normally used to define the higher-level responsibilities and reporting structure, but the explicit detail of information security responsibilities is normally in job descriptions, or other equivalent documents. Each employee having specific roles and responsibilities for information security should therefore have a document that defines their security roles and responsibilities, and other roles of which they need to be aware while carrying out their role (e.g. the incident management team). Auditors should check that this is all available to the relevant employees, and that employees are fully aware of their roles and responsibilities. One way of demonstrating this is to check whether the role/responsibilities document has been signed by both the employee and appropriate manager to signify understanding and acceptance.

Another aspect for the auditor to check is the date of the document, and whether it contains correct and consistent details relating to information security functions. A check of security responsibilities referenced in policy statements and individual procedures should provide full consistency with the responsibilities listed in the document (e.g. job description) held by the related role. Where security tasks and allocation of tasks have changed since the individual took on their current role, the auditor should look for either a reissued document or an addendum to the original one, again signed by the employee and relevant manager.

Organisations may vary in terms of where documents describing roles and responsibilities are held; some may be with the individual, and others with personnel/HR departments. In the latter case, the auditor should check that the individual has ready access to this information – they should have their own copy, as a person is unlikely to comply with a document last seen perhaps more than a year ago. Where individuals have jobs with specific security requirements, such as a network administrator, ensure that the job description or an additional document fully reflects this – general statements covering all employees are not acceptable in such cases. Individual roles with special responsibility for information security may also have explicit information security objectives set, with progress against those objectives recorded in regular performance reviews. Auditors should confirm that all documentation of this nature is current and properly controlled (see 5.9).

It is particularly important that those who are new to their jobs fully understand their responsibilities. The relevant training and actions (e.g. signing of document) to confirm both understanding and intent to comply with responsibilities should be completed at the time of appointment, before the new hire can accidentally contravene policy, not at the next convenient review.

Auditors should pay particular attention to temporary employees, contractors and third-party users. The same rules should apply to them; exceptions are not acceptable. There should be adequate descriptions of security roles and responsibilities for everyone working within the scope of the ISMS.

The clear definition and allocation of information security roles and responsibilities should be carefully investigated, as this can be a potential weak link. Overlaps in responsibility, without a specific person being given primary accountability for the activities taking place, may result in all parties in question leaving the matter to each other, and no one doing the activities or verifying that they have been done.

5.3 Segregation of duties (ISO/IEC 27001, A.5.3)

> *"Conflicting duties and conflicting areas of responsibility shall be segregated."*

Implementation guidance

Segregation of duties is a traditional business control used to reduce vulnerability to human error and deliberate misuse. Although most employees are essentially honest, there might be some who are not. Equally, honest people may be placed in a position where they will behave uncharacteristically (e.g. if under duress). A proportion of people may also become increasingly negligent over time if their activities are not limited. This can lead to problems with integrity (of people as well as information), loss of confidentiality, and resources becoming unavailable for their proper purpose.

Ensure that risk assessments properly identify and track the risks of unsegregated activities.

Dividing a process up between two or more staff provides a check at the point of handover, as one person can see that another has done what they were supposed to do. In sensitive areas, for example, the use of two keys or passwords by separate staff can ensure that no one obtains access to a resource without a second person either authorising or confirming their authority.

Many fraud and accounting deceptions are committed by single individuals who have been given access to too many functions within an accounting system, and/or individuals who do not need separate authorisation for their activities. A well-known disaster of this type was the Barings Bank incident. In 1995, a single trader, Nick Leeson, made a loss of £827 million before being detected. This resulted in the collapse of the entire business. The subsequent Bank of England report[3] indicated that this was caused by *"serious problems of controls and management failings within the Barings group"*.

Segregation prevents staff from operating on their own to create such incidents. Although the possibility of collusion remains, it is rare that two or more people will take the personal risk.

In small organisations, where segregation can be difficult to implement, the principle should be applied as far as possible with additional controls, such as increased monitoring, being

---

[3] *https://www.gov.uk/government/publications/report-into-the-collapse-of-barings-bank*.

implemented to compensate for any lack of segregation. If segregation of duties cannot be achieved, it can at least help to record all activities using a method that is resistant to tampering (also see 8.15), and to have procedures in place for independent review of these records to identify any suspicious or unauthorised activity after the fact; knowing that the records exist can act as a deterrent.

Auditing guidance

As noted in ISO/IEC 27002, 5.2 and above, small organisations may find it difficult to implement this control. This may be due to a lack of resources. At least for critical roles, some provisions should be in place – if nothing else is possible, crucial activities should be logged and the logs reviewed regularly by independent individuals for specific sequences of activities that are not permitted (also see 8.15). The auditor should ask to see these logs and evidence of the review process. It should also be shown that the individuals whose activities are being logged cannot alter or delete these logs, perhaps by the logs being stored in an unalterable format, then numbered sequentially by someone other than the subject of the logs and checked for missing entries as part of internal audit activities (see Clause 9.2).

For larger organisations, this control should be well established and properly demonstrated in their procedures. Of those procedures that should be considered for independent operations, security administration and audit are possibly the most critical and should be considered first.

The auditor should look at what independent verification of data and results is carried out between processing stages or before release. As part of its risk assessment, the organisation should have considered critical processes and whether any one person is responsible for carrying out too many of the checks and balances. Look at work arrangements for critical tasks: how are periods of sickness or holidays covered? Does this compromise independence? The organisation might need to enforce mandatory minimum annual holiday periods to achieve effective segregation. This approach is standard in some financial-sector organisations.

5.4 Management responsibilities (ISO/IEC 27001, A.5.4)

> *"Management shall require all personnel to apply information security in accordance with the established information security policy, topic-specific policies and procedures of the organization."*

Implementation guidance

A key success factor for all security programmes is support from management. Management roles are enabled to give this support by being aware of their duty to ensure that everybody in their area of responsibility (including themselves) is acting in compliance with approved policies and procedures, and is supporting/applying implemented controls. A list of typical management responsibilities is given in the implementation guidance of ISO/IEC 27002, 5.4. An important part of these responsibilities, in addition to the usual management functions, is to give employees, contractors and third-party users the message that their information security activities are recognised, valued and as important as other elements of their job function. The other key aspect is to execute the control function, and verify compliance with controls, policies and procedures during day-to-day work. This might not need complicated policing exercises – if management staff are aware of the correct application of controls, policies and procedures, they will easily detect if people in their area of responsibility are diverging from them.

Auditing guidance

The first thing auditors should check is that managers are aware of their responsibilities and understand their duty to ensure that employees, contractors and third-party users comply with controls, policies and procedures. The next step is to look for evidence that managers take this responsibility seriously – there could be several indications, such as mentioning information security in standing agendas for a meeting, or the reaction of employees when asked about management messaging regarding information security. Records of previous awareness evaluations (such as phishing tests) are also encouraging indicators. Another high-value indicator is whether employees feel that management is leading by example. Do individuals in senior roles comply with the same rules as apply to general staff (e.g. locking their computer when leaving their desk), or do they appear to have special dispensation to ignore the rules?

It can also help to ask about remedial training programmes for employees, contractors or third-party users whom management has identified as not complying with controls, policies or procedures, and what initiates such programmes. Another topic is what management does to motivate personnel, whether there are reward programmes for good suggestions, or other ways of actively encouraging personnel to support information security.

5.5 Contact with authorities (ISO/IEC 27001, A.5.5)

> *"The organization shall establish and maintain contact with relevant authorities."*

Implementation guidance

The organisation should have procedures in place to identify and establish all appropriate liaisons that must be in place with external regulatory bodies, service providers and any other organisation important for information security. In addition, the necessary approvals, non-disclosure agreements, reporting procedures and formats should be defined and approved to ensure that as much relevant information as possible can be exchanged.

It might be helpful to establish the function of a liaison officer who is responsible for contacting and liaising with authorities. This person can receive information from the authorities that might be helpful to guard against certain events, and to prepare for new legislation or regulations. They may also manage communications with authorities in the case of an incident (see 5.24 and 5.26).

Auditing guidance

This control requires appropriate relationships to be in place with external regulatory bodies, service providers and others that may have a crucial role in either preventing security incidents or mitigating their effects. The auditor should therefore look for evidence regarding the existence of the necessary contacts in business continuity and contingency plans, and infrastructure support documents. Contact details should be checked and updated on a periodic basis to ensure they are correct. For each contact, the purpose of the contact should be clear. Depending on the context, the auditor may ask the relevant liaison to make contact, to show that a chain of communication is established and to check that the contact details are accurate. The auditor can also check whether contacts are event-driven (e.g. following an incident) or time-driven (e.g. every quarter). Depending on the purpose of a contact, both types of relationship may be appropriate, but if contacts are really in use, the liaison should be very clear on how they are initiated and by whom.

The auditor should also look for evidence that legal, industrial, operational and technical requirements are being monitored for conformity as appropriate. The auditor should ensure that the organisation is able to demonstrate that it knows and has documented all applicable legal requirements, and that all contacts necessary to conform to these requirements are in place and up to date. The auditor should also review agreements and approvals to ensure that information being provided to relevant authorities is suitable, timely and authorised.

5.6 Contact with special interest groups (ISO/IEC 27001, A.5.6)

"The organization shall establish and maintain contact with special interest groups or other specialist security forums and professional associations."

Implementation guidance

Good ideas can be acquired by engaging with information security professionals from different organisations and sectors, many of whom have long and valuable experience in the subject, as well as by joining specialist groups, standards committees, etc. There are also closed fora that are devoted to a specific community or sector (e.g. public sector); these may require membership to be obtained via an endorsement or a vetting process. Although some bodies require a membership fee, they will usually give you a taste of what they have to offer before you decide. Other bodies do not have members but are useful sources of information. Various individuals have also made a name for themselves by providing interesting, accurate and timely information. Following their activity is an excellent way to maintain awareness of new threats, opportunities and trends.

Exchanges of security information should be controlled to ensure that confidential information is not passed to unauthorised persons. Some bodies operate on a strict non-disclosure basis to enable confidential discussion.

Auditing guidance

Evidence of suitable participation in specialist interest groups may be supplied via records of discussions on best practice and knowledge sharing on threats and approaches to manage them. A large organisation can be involved in security specialist groups, standards committees or similar activities outside its own environment. Smaller organisations are unlikely to be able to support extensive involvement, but attendance at appropriate free conferences and seminars would partly address this. In either case, auditors should check that procedures are in place to share and distribute any information received from such participation within the organisation in the most beneficial way. These procedures should also be designed to ensure that no confidential information is exchanged without proper authorisation.

5.7 Threat intelligence (ISO/IEC 27001, A.5.7)

> *"Information relating to information security threats shall be collected and analysed to produce threat intelligence."*

Implementation guidance

Threat intelligence can be obtained by research online, liaison with peers and procuring specialist services. In all cases, the work involves investigating threats to a sector, a geopolitical area or a specific organisation, putting the information into the context of the organisation to derive actionable insights, and maintaining that information in an accurate and sufficiently complete state (see Clause 4.1 of ISO/IEC 27001). It is vital to remain aware of the information being shared in (for example) online searches or provided to AI tools, as it may leak elements of the organisation's activities, risk posture and protections to potential attackers. These elements may then be pieced together by the attackers and used to improve their tactics.

Linking threat intelligence to risk management activities (see Clause 6.1.1 of ISO/IEC 27001) enables decisions on risks that may not yet have materialised but are likely to occur, with a greater level of reliability than predictions based on organisational data alone.

Auditing guidance

To identify that the organisation is obtaining and using threat information effectively to create threat intelligence, the auditor should interview staff with responsibility for making determinations on risks and selection of controls, to find out where they are obtaining the external evidence to support their decisions. Information security briefings and other updates to management may also include threat intelligence, and perhaps even a threat level such as is produced by some geopolitical entities to denote a national risk level. In all cases, the information on threats should be put into the context of the organisation ('what does this mean for us?') and referenced by management of the organisation in designing and running their activities relating to information security.

There should also be a demonstrable process supporting the repeatable production of intelligence, so that it is updated and remains as accurate as possible. A supporting process/standing guidance on how to verify or validate intelligence is a good sign of effectiveness in this area, as threat intelligence can vary dramatically in accuracy, even when provided by a trusted source. Finally, the organisation should provide suitable guidance and training to anyone tasked with direct threat intelligence research online, to comply with applicable legal and regulatory requirements, to protect the organisation, and to protect the individual carrying out the research.

5.8 Information security in project management (ISO/IEC 27001, A.5.8)

> *"Information security shall be integrated into project management."*

Implementation guidance

Projects are a major component of any organisation's work. Inappropriately controlled security during the delivery of a project can expose the organisation to uncontrolled business risks, resulting in information security incidents. Equally, the products and services that the project intends to deliver must also be suitably secure. A deliverable that provides what the customer asked for, but which also provides a route for compromise of an environment, is not fit for purpose. The chances of this outcome can be managed by integrating information security appropriately into project delivery.

Information security requirements should be recognised from the first stages of information systems acquisition or development, and all relevant requirements for information security should be specified along with the functional requirements. The development or acquisition of information systems should follow a well-specified and documented procedure that ensures that all identified security requirements are suitably addressed. The requirements analysis should refer to the results of risk assessments, and testing should be used to validate that the developed or purchased information system satisfies the identified requirements. ISO/IEC 27002, 5.8 contains a useful list of specific subjects to consider.

To ensure that a project preserves security during its lifespan, and produces secure deliverables, two types of risk assessment should be carried out at the beginning of the project. One type of risk assessment should consider risks relating to the activities of the project itself, and the other should consider risks relating to its deliverable(s). They should both produce security controls, which should be combined into one list and documented as part of project requirements. These requirements should then be prioritised according to the risks that the controls are intended to counter, and implemented as part of standard project activities.

Care should be taken to distinguish between the concept of 'project risk' and information security risk. The project may have a risk register that lists the risks to the delivery of the project, but this may not be a suitable vehicle within which to record and manage information security risk. Consider having an information security risk register, which integrates with the organisation's ISMS and is referenced explicitly from the project risk register. The information security risk register will then be less subject to the drivers that can overwhelm local project requirements. For example, a project may choose to remove operational requirements to improve its chances of delivering on time. Proposed changes to information security controls should be raised at the appropriate organisational level so that the risk to the organisation may be managed suitably.

Since projects inevitably change during their lifespan, information security should be reassessed as part of any project change process to ensure that risk mitigations are still suitable. Information security should also be considered at all key stages in the project's lifespan.

Finally, the transition phase of deliverables should be very carefully managed to ensure that project security requirements are realised appropriately in the deliverables that survive the project.

Auditing guidance

The auditor should first check the documentation for the project methodology in use, and check that it contains a requirement for identification of information security objectives, risks (as distinct from project risks) and controls. The methodology should also contain checkpoints throughout the project lifecycle to ensure that information security is addressed regularly. Any changes in the project should automatically initiate a review of existing information security risk assessment work to ascertain whether it needs to be revised. Changes should be made as appropriate to existing controls, new controls added, and/or controls removed. There should be a statement to the effect that responsibilities for information security must be defined and associated with particular roles.

In terms of what should be documented for each project, the auditor should seek evidence that there is a record of information security objectives and an information risk assessment for each project. They should also obtain evidence that information security risks can be traced to specific controls. Responsibilities for information security during each phase of the project should be defined, and roles should be associated with these responsibilities.

An organisation should be able to demonstrate to the auditor that information security requirements for the existing and new information systems have been identified and taken into account in the development and acquisition of applications, new systems, enhancements and upgrades to systems. This falls into two categories: where bespoke application software is developed specifically for or by the organisation; and where commercial off-the-shelf (COTS) software is acquired for use by the organisation. (Note that an insecure add-on component or customisation might compromise the entire application.)

The organisation's analysis of requirements should identify appropriate information security requirements for new systems (e.g. preservation of integrity within a given application), and these should be incorporated into the organisation's requirements documents. It is vital that this process takes place for all developed and purchased information systems. Having confirmed that this is the case, the auditor should check how these requirements are tracked, monitored and reviewed during system development or acquisition, testing, configuration and installation. The auditor should confirm that testing of the information system takes place, where necessary, to check that the requirements are met. Any identified deficiencies should be analysed, raised at the appropriate management level and satisfactorily resolved.

5.9 Inventory of information and other associated assets (ISO/IEC 27001, A.5.9)

> *"An inventory of information and other associated assets, including owners, shall be developed and maintained."*

Implementation guidance

An inventory of physical assets is required by accounting standards. For this reason, as well as for information security reasons, all organisations should have such an inventory. For information security purposes, the information assets that the organisation holds should also be included in an inventory. This may be in the same system or a separate one, as long as they are correlated appropriately (for example, enabling one to identify which servers hold which information). Appropriate protection can be applied to assets if it is known that the organisation has them, as then its information security, business and legal requirements can be assessed. The level of granularity of the asset inventory should take into account the requirements of the organisation, and the level of detail that will be necessary to support an effective risk assessment. Information can be grouped by business purpose, for example, perhaps as a list of databases or file names.

Possibly the most important concept regarding responsibility for, and protection of, assets is to assign asset ownership. It is necessary to appoint an asset owner for all major assets, and assets (such as personal data) to which special requirements are attached. This asset owner is responsible for the asset itself, and its protection. This includes:

- determining the classification of the asset (see 5.12);
- supporting risk assessments by giving input about the asset's business value and its importance for the organisation's business activities;
- ensuring appropriate protection in the day-to-day use of the asset; and
- keeping security classifications and control arrangements up to date.

See ISO/IEC 27002, 5.9 for further information on the activities of an asset owner.

It might be the case that the asset owner is not working with the asset on a day-to-day basis. In such cases, it is best that the asset owner appoints a custodian that works with the asset and looks after the asset on the asset owner's behalf. This custodian then looks after the protection of the asset in day-to-day business. It is important to note that the owner remains ultimately accountable and needs to verify that the custodian is taking their responsibilities seriously. As most organisations have many assets and/or complex systems, it can help to group several assets together, for example all assets involved in a particular process or in the provision of a particular service. The owner of that process or service may then be responsible for all the assets involved in the process or service, and for their correct functioning and protection.

It is important to note that some 'assets' may need to be retained by the organisation, and can constitute more of a liability than an asset. For example, identity documents that are required to be retained to prove that staff are authorised to work in the country. These assets cannot be used to generate revenue or positive value for the organisation. They should be treated as for other assets, with particular attention to their end of use, at which point they should be disposed of promptly.

For each instance of information and associated assets, the inventory should contain information about its business value and classification (see also 5.12), and its backup and disaster recovery arrangements.

The inventory of physical assets should also contain full details of equipment identity, including owner, location, maker, model, generic type (e.g. printer, PC), serial number, date of acquisition and inventory tag.

A record of physical and virtual disposals – when and how/who to – also needs to be kept, and the asset inventory should be updated whenever an item is disposed of. Organisational inventory tags (logo, inventory number) should be attached to all items that appear in the inventory. The organisation should ensure that all documentation (including system documentation), contracts, procedures and business recovery plans are included in the inventory; indicate the owner and those with operational responsibility. Information stored in non-technical formats (e.g. on paper) should be recorded, as well as that in technical formats (e.g. on CD).

All software products should also be listed in the asset inventory, including where they are used and how to obtain the reference copy (where relevant), together with licensing information. Adequate procedures should be in place to maintain accuracy of the inventory, including updating the inventory as part of change management and project activities. A 'stock check' should be carried out at least annually.

Auditing guidance

The auditor should confirm that the organisation maintains a complete and accurate asset inventory (or correlated physical asset and information inventories). This should include all major information (in whatever form, including software), physical assets, services and processes to be protected. The assessment will first need to determine that assets have been properly identified and classified (see also 5.12). The auditor should evaluate the inventory's adequacy.

- Is it sufficiently complete and accurate?
- Does it contain all necessary detail, and when and how is it updated?
- Are disposals recorded, when and to whom?

The auditor should confirm that owners have been assigned for all important assets, and that each owner is aware of the tasks and responsibilities that come with asset ownership. It might be helpful to review the records for asset classification and risk assessments to verify that the asset owner has participated in these activities to an appropriate level.

The auditor should also check the procedures in place to delegate routine tasks or the day-to-day use and protection of the asset to a custodian, and how, in case of such delegation, the owner checks that all tasks are carried out correctly.

As an asset inventory is only useful if it is up to date, the auditor should check the procedures in place to update the asset inventory, and how the introduction of new assets and the disposal of assets are reflected in the asset inventory. Project management and change management processes should be checked to see if they contain a step to update the asset inventory when changes occur that affect it.

The auditor should check that a role has been assigned responsibility for the asset inventory, and for its development and maintenance. They should also check how the inventory is protected. If the inventory is computer based, what about access control and backup? If paper based, where is it kept, how is it protected against loss, and what happens when the record is replaced? Are old copies kept? If yes, how long for and where? The asset inventory should identify:

- the item, format and, where applicable, its unique serial number, date, etc.;
- its business value and security classification;
- any special characteristics that may attract requirements (e.g. if personal data is involved);
- owner;
- location(s);
- media (if information);
- licence (software);
- retention period/lifespan;
- information about backups and disaster recovery; and
- date of entry and/or audit check.

5.10 Acceptable use of information and other associated assets (ISO/IEC 27001, A.5.10)

> *"Rules for the acceptable use and procedures for handling information and other associated assets shall be identified, documented and implemented."*

Implementation guidance

Every organisation is vulnerable to staff, and others, misusing assets, i.e. using them in a way that differs from their authorised business purpose. This may be unintentional or deliberate. In either case, misuse can impact the integrity of data and systems, threaten availability and expose confidential information. Additionally, information processing systems can be misused to attack other organisations.

A typical example is the Internet. This can be used for casual browsing during working time (which is possibly not an authorised business use) or to launch an attack from the organisation's network into other networks – an action the organisation could be held liable for.

In some circumstances, misuse of computers can be a criminal offence, for example in the UK under the Computer Misuse Act 1990. Similar legislation is in place in many other countries. The details of the applicable legislation should be checked to inform control selection.

To ensure that assets are only used for their intended business purpose, the organisation should identify, develop and implement rules, procedures and guidelines describing their acceptable use. These rules can vary considerably depending on the asset considered. The intended business use, past incidents and the owner of the asset can give valuable input when developing the rules for acceptable use. Any different use should be considered improper use, and all staff should be aware of that. The procedures should be adequate for the sensitivity level of the information handled, in line with the classification applied. It is also important that these procedures cover all sensitive information, regardless of its form.

It is important that everyone using the asset signs up to these rules, including not only the organisation's employees but also any contractors, third-party users or anyone else using the asset.

Auditing guidance

The auditor should confirm that acceptable use rules clearly describe the intended use of the assets, and the limits of this intended use. Where information classification is in use (see 5.12), the auditor should confirm that, for each classification label, there are procedures in place supporting information assigned that classification, such as procedures for secure processing, storage and transmission.

The organisation should apply controls to detect misuse. Disciplinary procedures should deal with actions upon discovering intentional misuse (see also 6.4) and inadvertent non-compliance. Investigate the use of other, peripheral or associated equipment such as printers, copiers, etc. What is the policy here? Is a suitably brief and readable warning message displayed at logon, making the user aware that unauthorised use of information processing facilities is not permitted?

The auditor should also confirm that asset users are aware of these rules. Evidence could be that the rules have been signed before access to the asset has been given. It is also important to talk to users to see if they retain an awareness of what constitutes inappropriate use. The auditor should review incident reports to see if they identify where the rules of acceptable use have not been followed, or have not functioned as intended. The reports should also lead to changes in the rules of methods for circulating them, as appropriate. If the organisation has no rules in place for the acceptable use of assets, there should be valid reasons for not doing so, and this should be supported by the findings of the risk assessment. Again, it might be helpful to look at incident reports to identify areas where rules for acceptable use would be helpful.

The auditor should confirm that procedures are in place to protect sensitive information – regardless of its form – in line with the classification scheme used by the organisation. Is it recorded? Who is responsible for this information? Who authorises its release? Who has received the information, and who is authorised to access it? Is clear labelling applied? Is distribution of sensitive information only taking place if there is a need to know?

Where information is being handled by persons unknown to staff – such as couriers – what additional identity checks are made? Are access restrictions in place, and if so, which ones? Check that agreements with external parties support the classification and handling procedures that internal staff are required to follow, and that they provide appropriate interpretations for the classifications. Checks of confidentiality or non-disclosure agreements also need to be made (see 6.6).

Observe how people in the organisation are handling sensitive, critical and personally identifiable information, and how easy or difficult it might be to circumvent or disregard the procedures.

5.11 Return of assets (ISO/IEC 27001, A.5.11)

> *"Personnel and other interested parties as appropriate shall return all the organization's assets in their possession upon change or termination of their employment, contract or agreement."*

Implementation guidance

The organisation should have procedures in place to ensure that all assets in the possession of employees, contractors and third-party users are returned when their employment terminates or changes (see also 6.5). It is important that this covers all assets: not only equipment and software, but also all the organisation's documents and any information stored on media. Depending on the organisation, its business and the particular job function, there might also be other assets, e.g. credit cards, access cards, manuals and mobile devices.

All the organisation's information that might have been stored on non-organisational assets, such as private equipment, or equipment of a third-party organisation or of a contractor, should also be returned and securely erased from that equipment. The return of assets should be part of the contract.

Auditing guidance

Auditors should examine the organisation's procedures to ensure the return of assets and transfer of knowledge. There are several issues that these procedures should address.

- They should cover all assets, ranging from hardware and equipment to information in any form (electronic, paper, on storage media, slides, films, etc.), and should also cover keys or cards that are used for access control purposes.
- They should cover the transfer of any information stored or processed on non-organisational equipment or media, and the secure erasure of the information from the equipment or media.
- They should be applicable to employees, contractors and third-party users.

The auditor should also check that these procedures are not only applied upon termination, but are also applied if employment changes and the assets are no longer required in a new job function. Records for each leaving or moving employee, contractor and third-party user should verify the return of assets.

5.12 Classification of information (ISO/IEC 27001, A.5.12)

"Information shall be classified according to the information security needs of the organization based on confidentiality, integrity, availability and relevant interested party requirements."

Implementation guidance

Information with different requirements for confidentiality, integrity and availability will require different treatment. To simplify this, it is common to group information with similar requirements for protection. The process of assigning a specific piece of information to a specific group is known as 'classifying' it, and the group to which it is assigned is known as its 'classification'. For example, a medical record could be classified as 'highly confidential'. There are different classification schemes in use by governments and large and small organisations, and often the same classification term (e.g. 'company confidential') may have different meanings in different organisations – and sometimes even within a single organisation, in the case where a client requires its data to be classified using its own scheme. There is no international standard for the classification of information. It is not usual to classify anything other than information – but it is possible to specify, with regard to another sort of asset (e.g. a server), what classification(s) of information it is permitted to handle.

There should be a clear initial decision as to whether all information must be actively classified, or whether the act of classification is optional, and what that means (e.g. all information automatically has the classification 'Company Internal' unless otherwise specified).

The classification scheme should be in writing and available to all those with authority to apply it, i.e. all those who originate documents and data. It should also be available to those who need to understand what the classification means when applied to information. The classification scheme should be easy to understand and clearly differentiate between levels to support the correct assignment of classification levels. Too many levels will lead to inconsistency as staff cannot differentiate between definitions. Too few levels, and staff will find that they need to over- or under-classify.

The handling, storage and disposal requirements of each classification should be specified, and supporting procedures created. Allowance should be made for the need to periodically review classifications that have been assigned, and to change the level of classification when/if the sensitivity changes. Provide change and expiry dates to record these. The asset owner should be responsible for classifying their asset(s), ensuring that handling rules are complied with and updating the classification as appropriate (see also 5.37).

Auditing guidance

Auditors should confirm that the organisation has developed or selected, and implemented, an adequate and consistent classification scheme, supported by training and handling rules. For assets to be properly protected, there should be some form of grading based on their individual requirements for confidentiality, integrity and availability. The classification that is assigned to an asset, and the associated handling requirements, should take account of business requirements for exchanging and sharing information, as well as the security requirements of the asset. The classification scheme should be applied to all assets considered in the scope of the ISMS. Without clear classification, assets may not be properly protected. The scheme should not be too complex and should be supported

by arrangements with other organisations to ensure that the
interactions between different classification schemes are
understood. Do the procedures account for how the correct
classification is to be checked? Does a procedure to review
and upgrade or downgrade the classification level exist?

The auditor should confirm that the classification and
handling scheme is readily accessible, understood by all staff
and regularly reviewed. The owner of an asset should be
responsible for assigning its classification,
applying/supervising appropriate handling and updating the
classification if anything changes.

5.13 Labelling of information (ISO/IEC 27001, A.5.13)

> *"An appropriate set of procedures for information
> labelling shall be developed and implemented in
> accordance with the information classification scheme
> adopted by the organization."*

Implementation guidance

All information assets should (as appropriate – see below) be
prominently labelled in a manner suiting their form to ensure
that they can be associated with the necessary handling
guidance in use, storage and transport (see 5.10). All printed
and bound documents should contain the appropriate
classification label (unless 'unclassified' is a meaningful
concept in the organisation; see 5.12), and loose-leaf
documents should carry it on every page.

Information held in online/digital form should also be
classified, although it is sometimes challenging to label it;
labelling may happen at the document level, or at a higher

level (folder/directory), or even at a volume level, through associated policies (e.g. 'all data in the Cloud 3 Environment is classified at OFFICIAL'). Some security systems may also include a security labelling facility. Handling requirements should be supported by appropriate technical controls, such as user access management (see 5.15). In any case, the organisation should ensure that any information in transit from a given location (e.g. by email), or that has changed in form (e.g. a printout) retains its label.

One consideration that may significantly affect this advice is where it is inadvisable to make it obvious which information is the most valuable or sensitive, as this essentially provides a shopping list for an attacker. In this case, only roles that need to know of the most sensitive information should be aware of its existence, and labelling may be disguised. This, of course, may lead to the risk of inadvertent mishandling should the information fall into the wrong hands, so the control of distribution of this information, and the thorough training of roles that handle it, is vital.

Care should be taken in interpreting classification labels on documents from other organisations because different organisations may have different definitions for the same (or similar sounding) label. Equally, ensure that your classifications will be properly respected when sent to other organisations.

Information may cease to be sensitive after a certain time, for example when it has been made public. In such cases, provide an expiry date to avoid unnecessary protection expense.

Auditing guidance

The auditor should confirm that the organisation has procedures for the labelling of classified information, compatible with the classification scheme. Auditors should also confirm that the marking suitably represents the most sensitive item in the entity (e.g. an information processing system or a database).

Labelling physical items, such as documents, tapes, hardware, etc., is straightforward, but what about information held in information systems, and correspondence electronically transferred? The auditor should confirm that the solutions the organisation has chosen for labelling electronic formats have been checked for adequacy. Is this clear and understandable? Does it convey the correct label to the receiver of the information, and does this subsequently lead to sufficiently secure access, use or storage of that information? Are the labels of physical assets appropriate? Labels may be hard to find where they should be prominent; adhesive labels can become detached and leave the item unmarked and unprotected. The auditor should also verify that labels remain with information when it changes form (e.g. when printed out).

5.14 Information transfer (ISO/IEC 27001, A.5.14)

"Information transfer rules, procedures, or agreements shall be in place for all types of transfer facilities within the organization and between the organization and other parties."

Implementation guidance

The exchange of information, in whatever form, carries many risks of compromise. The organisation should have an information exchange policy and supporting procedures in place describing the rules to be applied when exchanging information, developed to support the expected and likely methods that will be used. When creating this policy and the supporting procedures, the items listed in ISO/IEC 27002, 5.14 should be considered.

The policy should be communicated to all employees, and awareness training with real-life examples used to illustrate the risks involved. All personnel should also be aware of the possibilities of:

- information being leaked if sent to the wrong address (physical or virtual);
- information being intercepted by unauthorised people if sent unprotected;
- information in paper form being compromised when left unattended in printers or fax machines;
- staff being overheard when using mobile phones in public places such as trains; and
- the wrong person picking up a fax or listening to an answer machine message, despite the right number being dialled.

Email and other forms of electronic messaging are complex and have multiple levels at which an attacker could compromise the information being exchanged. The controls in place to protect them should therefore be carefully selected to protect messages against deliberate or accidental loss, damage/alteration and exposure. Internal messages could inadvertently be sent to external parties, contracts can be implied, and messages and attachments could be retained on backups past their authorised retention period. Standard disclaimers attached to electronic messages might help, but their legal status as protection is not clear.

Agreements or contracts with a third party should explicitly specify the controls to be applied by the organisation and the third party when exchanging sensitive information. Agreements should be authorised at an appropriate level in the organisation and periodically reviewed. Changes in practice should always be change controlled and reflected where necessary in agreements.

Auditing guidance

The auditor should check that the organisation has a policy for information exchange and supporting procedures, and that this policy and supporting procedures address the different types of exchange that the organisation uses. The auditor should also check, either through reviewing procedures or through technical means, that information exchange violating the policy is not permitted. The policy and supporting procedures should cover all forms of communication facilities, including networks, mobile computing devices, telephones and mobiles, fax machines and answering machines.

Auditors should review procedures against the items described in ISO/IEC 27002, 5.14. The auditor should confirm that employees are aware of the procedures – for example, by asking them about their use of email, SMS or messaging apps to find out whether they are aware of the risks involved and how/whether they should be using these services.

Auditors should:

- check which third-party organisations are involved in the transfer of sensitive information;
- confirm that the necessary contractual documents exist; and
- verify that these documents address the correct treatment of sensitive and confidential information in transit.

The auditor should check what the organisation has done to investigate the controls that its suppliers, customers and partners use when communicating with it, and investigate how the organisation has addressed situations where third-party protection requirements and controls vary from those that the organisation has determined appropriate. The auditor should also verify that agreements between organisations exchanging sensitive information or software are in place.

The auditor should check the organisation's security arrangements for electronic messaging, asking questions such as: How is information included in and attached to electronic messages controlled? How is incoming data verified as to source and integrity? If applicable, what encryption methods are applied (also see 8.24)? Is information received from electronic messages checked for malware before use? What services are permitted for what sort of data?

Auditors should check the organisation's security arrangements if use of external messaging platforms is permitted. Access to a specific messaging platform may be restricted to certain roles or use cases. If so, how is this monitored and enforced?

5.15 Access control (ISO/IEC 27001, A.5.15)

> *"Rules to control physical and logical access to information and other associated assets shall be established and implemented based on business and information security requirements."*

Implementation guidance

Access control is a fundamental prerequisite to securing information or services on information processing systems, and is also necessary to protect physical premises containing information in all forms. The organisation's business information, together with other business assets (such as its reputation), are at stake.

Allocation of user access permissions should be driven by business requirements, approved and regularly reviewed by asset owners, and clearly stated in the access control policy. Any access not necessary for the activities assigned to a role should be denied. This approach is known as 'least privilege'.

The organisation should develop, document and implement an access control policy that defines user access rights based on business needs, considering the classification and handling requirements (see 5.12) of the information, services, networks and/or applications accessed, and any legal or regulatory requirements. Standard user access profiles for specific roles are a useful way of managing access where many users are involved (this is known as 'role-based access'). It is important to review the access control policy regularly and remove any access rights that are no longer necessary. Note that business information repositories such as calendaring systems should also be taken into account.

Good change control and management is essential to keep access rights up to date, and regular monitoring is required to provide assurance.

Auditing guidance

Auditors should confirm that access control rights and rules are clearly defined in the access control policy, and that they are consistent with the classification and handling requirements for the information. Suitable mechanisms for enforcement of this policy should be in place and implemented, for example system and network access controls. An access right to an asset should be traceable to a risk assessment and authorised by the correct asset owner (potentially via a grouping of rights, not per individual asset). Any access to sensitive information (also see 5.12), or to information processing facilities, should be based on the 'need to know' and 'need to use' principles, i.e. justified by business requirements and necessary for the task at hand. Role-based access should be implemented where possible. Users accessing a resource (e.g. an application) should be provided with access only to the information and services that they have been authorised to use.

Auditors should be prepared to question why certain roles, especially senior ones, have access to certain information if this access is not sufficiently justified. Check also that access to sensitive information takes place in line with the classification given, and that the access permissions given have been reviewed to ensure that they are consistent with applicable legislation and regulations. Also check that personnel with access to sensitive or confidential information have been properly trained, since unrestricted use of such information by untrained staff can have disastrous consequences.

The auditor should also check that the organisation has procedures in place to review the access control in place, taking account of employees leaving the organisation, job functions and requirements changing, etc. Records should exist to show that access permissions are promptly updated to reflect the organisation's current requirements.

5.16 Identity management (ISO/IEC 27001, A.5.16)

"The full life cycle of identities shall be managed."

Implementation guidance

The organisation should formally and uniquely register every user, and maintain a record of each information system, network or service that they have a business requirement to access. Failure to control registration can result in breach of confidentiality, unauthorised modification and/or loss.

The sharing of user identities almost guarantees loss of accountability, as actions cannot be unambiguously traced to individuals. This can then lead to loss of confidentiality, integrity and availability. Users should therefore each have a unique identifier, and hence identity, for every system they have authorisation to access.

In those few cases where it is not possible to have individual identities (e.g. because the system does not have the required functionality), the organisation should implement a manual process to track who is using the identity at any given time, ensure that it cannot be used by more than one person at any time and change authentication credentials whenever the identity is passed to a new custodian.

Auditing guidance

The auditor should collect and check relevant evidence to be assured that the user registration and de-registration process functions effectively. This may include records of starters, movers and leavers, process documentation, and lists of authorisations for systems. The auditor should also check that system privileges match documented authorisations, and that user identities on the systems match registered users.

The term 'user' should be taken to include all users of information processing facilities, including system administrators, managers, application users, technical support personnel and programmers.

The process for creating and removing user identities should be documented and logged, and the process for employees leaving the organisation should include prompt removal of user identities. There should also be a process for auditing live identities on a regular basis to catch any that may have been inadvertently left live when they are no longer required. Redundant user identities should not be reissued to other users because of the risk of inadvertently giving excessive and unauthorised access to resources.

The auditor should check whether user identities are unique or shared. If shared, why is this necessary? Look at risk assessments, and the management and authorisation for this. Are additional controls applied to provide accountability, and are these additional controls sufficient?

5.17 Authentication information (ISO/IEC 27001, A.5.17)

> *"Allocation and management of authentication information shall be controlled by a management process, including advising personnel on appropriate handling of authentication information."*

Implementation guidance

User identification and authentication goes hand in hand with access control, user registration and allocation of privileges. This is very often done through passwords, which remain a common, but not the only, mechanism for user authentication. Secret authentication information also includes biometric data and PINs, but not user IDs or email addresses (as these are not secret). It also includes the answers to questions that are commonly used to allow a person to do a self-service reset of their password. Exposed (written down) secret authentication information, or obvious and easily guessed passwords, can lead to misuse of systems by unauthorised persons, with the attendant risks of loss of confidentiality, integrity and availability.

Whatever process is used to allocate secret authentication information, it should be based on the positive identification of the user, and a formal process applied enforcing users to change initial temporary passwords. The user should acknowledge receipt of their credentials.

A procedure is required to ensure that user IDs and passwords are issued only to those with a business need for access (see 5.15), and are authorised properly by the owner of the resource being accessed. Where other methods of user authentication are being used, similar controls will be required but potentially with additional controls suitable for the method employed.

Users should also be made aware if they may be held accountable for actions carried out by someone else having used their sensitive authentication information.

Weak password management leads to the risk of misuse of systems by unauthorised persons, with the consequent risks of breaches of confidentiality, integrity and availability of information. Where systems can be used to enforce password quality, lifespan and frequency of change, these should be used. If this is not the case, compensating controls and monitoring may be required. Use risk assessment to determine whether this is necessary.

ISO/IEC 27002, 5.17 lists guidelines for good password management systems, password management and related user responsibilities in controls.

Auditing guidance

The auditor should check that there is a policy covering the use of passwords throughout the organisation, including ensuring that a suitably secure password management system is used. The policy should be consistent with the security requirements of the organisation and the information protected. Any requirements for particularly sensitive areas should be additional to, and consistent with, this policy.

Aspects that auditors should check for include:

- length and content of passwords;
- frequency of changing passwords (both maximum and minimum limits);
- use of individual user IDs;
- use of common passwords between individuals and/or by the same individual across different applications;
- secure handling and storage of passwords; and
- changing of default passwords.

Additional desirable attributes of a password management system are given in ISO/IEC 27002, 5.17. Some systems enforce rules of this type; others do not. Are the technical rules used to verify sensitive authentication information consistent with documented policy? Check what other measures are in place if the system does not automatically enforce such rules.

The auditor should check how secret authentication information is initially allocated and controlled. In some cases, allocation may be done by the user, but not in all cases – initial passwords may be randomly generated, for example. If centrally controlled, where is the information held? Is it secure? Who has access? If allocation is under user control, are they aware of their responsibilities (see also 5.2)? Are procedures in place to ensure that temporary or default vendor passwords are changed immediately? If appropriate, ask staff to show you how they change their sensitive authentication information.

Auditors need to review records to verify that users have signed a statement to keep their secret authentication information confidential, and interview a randomly selected group of users to check that they are aware of and understand this responsibility. This awareness should be tested at different levels of the organisation. There should also be a record of how each user's identity was verified at the time of provision of the information, as well as a list of acceptable forms of ID. Look for records showing that initial secret authentication information has been changed by the user on first use. Ask users what happens when they forget their passwords. How are new passwords provided? Is there an additional route for resetting passwords if the matter is deemed urgent? Are there follow-up processes to review the use of an urgent password reset (see 5.29)?

The auditor should also note any instances of sensitive authentication information being written down on memos, stuck to monitors, etc. It is also instructive to ask for the results of password-cracking tools that the organisation might have used to check the quality of user passwords. Auditors should ensure that proper management authority is obtained when investigating the use of sensitive authentication information.

5.18 Access rights (ISO/IEC 27001, A.5.18)

> *"Access rights to information and other associated assets shall be provisioned, reviewed, modified and removed in accordance with the organization's topic-specific policy on and rules for access control."*

Implementation guidance

Inappropriate allocation of access rights increases the chance of deliberate or accidental misuse.

Hence access rights should always be based on business needs. However, with the best will in the world, it is possible that a mistake may be made, and access inappropriately granted, or preserved when a user leaves or changes role. The need for access should therefore be reviewed periodically, and access rights should be withdrawn at that point if it is found that they are not needed. This is particularly important where users have access to sensitive information or have elevated privileges. In cases where users use a shared password to access a resource, it will be necessary to change that password when any of them no longer require access to the related resource; individual credentials may be a preferable option.

A user registration form should be prepared, on which the information system, network, service or application(s) required is described, as well as the conditions of access. This should be signed by the applicant to document their acceptance of the conditions, and by the system owner or custodian to document their authorisation for the applicant to be registered. This form should have the user ID added to it and then be archived.

It is important that user access to resources is promptly disabled when someone ceases to have a business reason to access the resources, for example on termination of employment or internal job move. Procedures should be put in place to ensure this. There should be notification procedures, and clearly defined responsibilities and actions when employees leave the organisation or change their role. This should take into account physical access rights (i.e.

returning of access control cards, keys, identification cards, etc.; see also 7.10), as well as rights to log in to the organisation's network, user accounts, passwords, email addresses and any other form of permitted access. It will also be necessary to review and update the documentation listing the individuals who have access rights, are assigned to a role, any subscriptions that are tied to them and memberships in interest groups.

Role-based access control should be considered, as this can be easier to manage, and reduces the chances of 'special cases'; or, if they do appear, it makes them more visible and therefore harder to authorise without adequate justification.

Special procedures might be required if termination of employment is initiated by management and/or relates to disaffected employees, contractors or third-party users, to avoid information collection and disclosure, modification or destruction of information or services in the period between the individual becoming aware that their employment is terminating and the point at which they lose access to information and information systems.

Auditing guidance

The auditor should collect and check records of registrations and authorisations, and check that user access levels are based on formal registration and authorisation of the users, and recorded. The auditor should also check that these records are consistent with implementation. Have staff who have moved away or changed to other responsibilities had their authorisations immediately removed from the appropriate systems? Interview staff who have changed role – have they retained previous privileges that they no longer need? Those interviewed should include staff members who have been with the organisation longest, and those who have been promoted within the same function.

Another aspect that the auditor should check is whether users who change role are promptly granted any necessary additional access. If this is not done, users will tend to start sharing IDs to enable them to get their jobs done. Where the addition of required privileges is frequently delayed, this temporary credential sharing may be a widely accepted workaround that has not been risk assessed. If a shared credential is an authorised approach, then the auditor should verify that it was changed last time one of the users with shared access ceased to require access (e.g. because they left).

Auditors should spend time with the system administrators looking at operating system settings for access control of specific groups and individuals, ensuring that access can only take place for registered and authorised users. ISO/IEC 27002, 5.16 gives further information on managing user IDs. It is also worthwhile checking that users are aware of their access rights and restrictions, and understand that they should not try to circumvent access controls.

The auditor should check that procedures for regular review of all kinds of user access rights and privileges are in place and are being followed. These procedures might be a formal audit to check compliance, followed by a management-level review to check for consistency with business and policy requirements. The auditor should check that reviews are logged, that actual access has been compared to authorised access, and that the authoriser for the system has verified that authorised access is appropriate. Changes should be reviewed on a periodic basis.

The auditor should check that the organisation has procedures in place that define all actions to be taken to remove access rights in case of employment termination. These procedures should be applied in case of any termination, irrespective of whether this relates to an employee, a contractor, a third-party user or anyone else having had access to the organisation's premises or assets. The access rights considered for removal should include all physical and logical access, access to services and involvement in user or interest groups. They should cover the notification of staff and other relevant parties of the departure of the user, with instructions to cease to share information with the user that the user should no longer have access to.

5.19 Information security in supplier relationships (ISO/IEC 27001, A.5.19)

> *"Processes and procedures shall be defined and implemented to manage the information security risks associated with the use of supplier's products or services."*

Implementation guidance

There are several ways suppliers can cause risks to the organisation's information and information processing facilities. This might be via physical access as well as via logical access, for example using online connections or remote working with the organisation's assets. It can also be by incidental exposure to sensitive material, such as papers on desks while someone is watering plants in an office.

In addition to the above, the use of a supplier's services may include provision of sensitive information to that supplier, or use of information generated by the supplier to make business-critical decisions. So even if a supplier has no access to the organisation's systems, it may still present a significant source of risk.

If the organisation intends to use supplier information to assist its decisions, or allow suppliers to have access to sensitive data or to secure environments, it needs to have an overarching policy covering what suppliers must and must not do. It should also have a master list of suppliers. The risks that apply to each working relationship with suppliers need to be identified and assessed.

To manage the implementation of this policy, the organisation should create a standard process and procedure for managing supplier security, tailored to the security level of the information to which the supplier will be exposed, and the impact of the supplier's activities and information on the organisation. This procedure should take into account that the supplier may itself have other parties to which it has delegated certain activities.

Auditing guidance

The auditor should confirm that there are a set of principles and a top-level policy and procedure for handling supplier security. ISO/IEC 27002, 5.19 contains a list of the types of information that can be in the policy. The auditor should also check that risk assessments are carried out to assess the risks related to suppliers.

Ask for a master list of suppliers. This should indicate the security level of the information to which each supplier is authorised to be exposed, and should reference the specific agreements made with that supplier (see 5.20).

5.20 Addressing information security within supplier agreements (ISO/IEC 27001, A.5.20)

> *"Relevant information security requirements shall be established and agreed with each supplier based on the type of supplier relationship."*

Implementation guidance

The same level of security as that which applies to the organisation's staff should be applied to supplier staff who are able to access the organisation's physical or logical environments, including user IDs, passwords, data access controls, physical security, etc. What needs to be taken into account when developing the agreement that regulates supplier access is that the organisation does not have direct control of the supplier's management, personnel controls, IT, and security policies and practices. The supplier may also have a different risk appetite and business practices. These differences should be identified and assessed as part of due diligence when determining whether to work with the other party.

The key document that needs to be in place before any sharing of information or access is a contract or an agreement. It should provide details on the facilities that each party will make available to the other, and the security controls to be put in place, as well as which entity is responsible for which security controls. Suppliers should not be given access to the organisation's information and/or information processing facilities until the appropriate controls have been implemented.

The implementation guidance in ISO/IEC 27002, 5.20 provides a list of suggested items to put in place as required by the results of the risk assessment. The contract or agreement clauses may also specify conformance with ISO/IEC 27001, or even certification, again depending on the requirements. Ensure that the signatories on both sides are properly identified and authorised.

The security documentation should include copies of all relevant contracts or agreements, and possibly several additional documents describing specific elements of the relationship. It might be helpful to include security controls, policies and procedures in a security plan that can be given to the third party. Any deviation from these requirements should be justified and documented.

Auditing guidance

The auditor can check the results of the risk assessments that the organisation has carried out for its suppliers. Risks can come from remote access to mainframe or server software, the Internet connection, badly isolated intranets or physical access to secure areas.

The auditor can also check that all security requirements and risks for supplier arrangements are identified and addressed in a formal contract or service level agreement between the two organisations. The implementation guidance in ISO/IEC 27002, 5.20 provides a list of issues that should be considered for inclusion in such agreements. The auditor can check that the organisation has adequate procedures in place to ensure that all security issues are addressed before giving supplier access to any of the organisation's assets.

It is also necessary to ensure that suppliers are aware of all the security arrangements they have to put in place, and understand and agree to these arrangements. An approach could be to ask the supplier for its ISMS certification. The auditor could also ask the organisation how the agreements cover the situation where a supplier does not perform in accordance with the expectations of the organisation. The auditor can check that all relevant liabilities and potential disruptions have been identified and are addressed appropriately. Provisions need to be in place for modifying agreements, when necessary, as well as for their termination.

5.21 Managing information security in the information and communication technology (ICT) supply chain (ISO/IEC 27001, A.5.21)

> *"Processes and procedures shall be defined and implemented to manage the information security risks associated with the ICT products and services supply chain."*

Implementation guidance

Having a relationship with an ICT supplier (e.g. an application vendor or a cloud hosting provider) usually means that, through them, the organisation is exposed to risks posed by other suppliers to that first-party supplier. For example, the vendor of a piece of technology may outsource maintenance to another company that operates as a franchise.

When specialist advice is required, the franchise staff may then bring in staff from the manufacturer. The manufacturer will have accounting and HR software, as well as possibly be using cloud services for storage of customer data – the 'supply chain' is really a multiply connected network. Each of these entities may be able to make mistakes or carry out deliberate actions that affect the organisation's security.

This network of suppliers should be managed via agreements with the supplier(s) with which the organisation has a direct contractual relationship. ISO/IEC 27001, A.5.21 contains a list of considerations when designing a supplier agreement to manage this 'inherited' risk. The organisation should also actively investigate its ICT supply network to a level of detail commensurate with the risk, and identify any increased areas of concern.

One key area is how these suppliers change, and how they are chosen. Each entity in the network may do this differently, producing possible gaps in security.

Auditing guidance

The auditor can ask to review supplier agreements, and compare them against the list in ISO/IEC 27001, A.5.21. Are all relevant topics taken into account? Has a risk assessment been carried out on all supply chains? Have appropriate controls been identified and implemented? How is their effectiveness assessed? The auditor can look for suppliers whose activities may impact information security indirectly, such as those related to disposal of sensitive waste, server room maintenance or laboratory cleaning.

5.22 Monitoring, review and change management of supplier services (ISO/IEC 27001, A.5.22)

> *"The organization shall regularly monitor, review, evaluate and manage change in supplier information security practices and service delivery."*

Implementation guidance

Once the operations of the service provider have begun, it is up to the organisation to ensure that the services delivered conform to the requirements specified in the contract. One of the most important means to ensure that security controls, service definitions and service delivery levels are being provided as specified by the supplier (see 5.19) is to monitor and review these controls and services. This can include checking obvious issues, such as the availability levels of the provided service, or it can be something more involved, such as direct or indirect verification of the technical security controls in the supplier's environment. The organisation should put procedures in place to monitor service delivery and to review service reports produced by the supplier. ISO/IEC 27002, 5.22 provides a list of activities that the supplier service management process should include.

A list of possible types of changes in supplier services is given in ISO/IEC 27002, 5.22. The organisation should have procedures in place to respond to, and actively manage, these changes. This includes reassessing the risks involved in the new/changed arrangements, negotiating variations to changes, and modifying supplier contracts or entering into new agreements or contracts. It is important that this takes place following a well-defined procedure, and with appropriate authorisation.

Particular care should be taken in the case of changes that alter risk levels. If something has changed within the supplier, or in the technology used to provide the services, the organisation should initiate a review of the security controls in place, and make changes as required to maintain information security risk at a level acceptable to the organisation. Requirements imposed on the organisation by other parties, such as customers, should also be reviewed to ensure that they are still being honoured after the change.

Another issue to be managed is how the supplier responds to information security incidents. The contract or agreement should state that the supplier provides the organisation with notifications about information security incidents, and the organisation should have assigned responsibilities, sufficient resources and procedures in place to review these reports, and initiate its own incident management process if required. Incident management reports should also be reviewed to verify that the supplier addresses these events adequately. The same is true for any other problems, faults, events, etc. that could have an impact on the organisation's information or the services provided to the organisation.

If the agreement or contract allows audits to take place, then this is another way the organisation can verify that the supplier acts in accordance with the contract or agreement.

Auditing guidance

The auditor can look for evidence that the organisation is receiving all relevant reports, records and logs of services provided, and that it has procedures in place to review these.

To check this, the auditor can ask for records or reports, and for any records that are produced as a result of the reviewing activity of the organisation. These documents should also show that relevant activities, for example as listed in ISO/IEC 27002, 5.22, are taking place.

In addition, the auditor can look at records of actions the organisation has carried out to check the security controls, service definitions and service delivery levels to ensure the procedures are applied correctly. The organisation should also have procedures in place to react to any nonconformity of the supplier with the requirements specified in the contract or agreement.

The organisation should have procedures in place to manage any changes to services provided by suppliers. The auditor should check that these procedures include reassessing the risks, taking account of the possibly changed business requirements and the systems involved. The auditor should also check that the process requires management approval before any changes are made, and that all relevant stakeholders, for example roles with responsibility for legal matters, are given the opportunity to review changes to the contract or agreement.

The auditor can also confirm that the organisation has assigned responsibilities for monitoring and reviewing, and that the people carrying out reviewing have sufficient skills and time available to carry out their reviewing activities.

If the agreement or contract allows the organisation to carry out audits of the supplier, the auditor should establish, for example by looking at audit reports, that the organisation does audit the supplier, and that any findings are actively communicated to the supplier, followed up and resolved satisfactorily.

5.23 Information security for use of cloud services (ISO/IEC 27001, A.5.23)

> *"Processes for acquisition, use, management and exit from cloud services shall be established in accordance with the organization's information security requirements."*

Implementation guidance

Where the organisation is using cloud services, these should be treated in the same way as for other third-party suppliers, with some specific areas of focus to take into consideration the special characteristics of cloud provision. for example, the geographic location of information is often very fluid in the case of major cloud providers; this provides a significant advantage in terms of resilience, in that the loss of a single location may have minimal effect on the availability of a service. However, the location of information determines the laws and regulations that apply to it, creating a significant challenge to any organisation faced with the answer 'it depends' when asking 'where do you keep our data?'. Many cloud providers have taken this factor into consideration when designing their services, so that location can be constrained. in more general terms, the organisation should always ensure that the cloud service that is implemented includes the precise criteria that meet its requirements, as features such as encryption, constrained location of data and constrained location of support teams usually carry additional cost. ISO/IEC 27002, 5.23 provides a list of types of criteria that should be included.

Cloud services are frequently highly variable in relation to what information security is provided by the third party, versus what is expected that the customer will provide; it is vital for all parties (especially where this is a three- or four-entity relationship) to have a clearly defined set of responsibilities and jurisdictions. There is also often an assumption that a cloud service provider will be the eternal solution – but what if the organisation wants to change vendor, or the cloud service provider unexpectedly goes out of business? when engaging with a cloud service provider, the exit strategy is as important as the engagement strategy.

Auditing guidance

The auditor reviewing cloud services should expect, in many cases, that the organisation has had to accept standard vendor security controls, which have resulted in a risk profile not matched to its risk appetite. They should verify that any risks above tolerance have been understood and accepted by the correct level of management in the organisation.

The lists of criteria and considerations in ISO/IEC 27002, 5.23 lay out the main areas that the auditor should look for in cloud service provider agreements. it is important to note that, in addition to these agreements being likely to be generic, they may also be complex and lengthy, and the auditor should seek assistance from relevant legal experts where appropriate.

With regard to ascertaining the level of contact between the organisation and its cloud service providers, the auditor should look for records of service review meetings and performance reviews. where these do not exist (for example where a small client is consuming standard services from a global cloud provider), there should be a dashboard or standard reporting process by which automated service performance data is shared with the organisation. Incident notifications should, in all cases, be provided proactively – it should not be necessary for the organisation to visit a portal to find out that an incident has taken place. Where incidents have taken place, there should be records of advice and guidance from the cloud service provider, and evidence that the organisation has received, reviewed and acted upon this advice.

5.24 Information security incident management planning and preparation (ISO/IEC 27001, A.5.24)

"The organization shall plan and prepare for managing information security incidents by defining, establishing and communicating information security incident management processes, roles and responsibilities."

Implementation guidance

Information security incidents can result in breaches of confidentiality, failure of integrity of equipment and data, and loss of availability. They provide a valuable opportunity to improve procedures and processes to prevent them occurring again, or to ensure that they cause acceptable levels of impact when they do recur. Examples include fire or flood, electrical failure, hardware breakdown, failed software, virus infection, unauthorised access (actual or attempted) to controlled premises or to computer systems, corrupted or lost data, misdirected emails and failure of any security control.

The organisation should have procedures in place to ensure an orderly and effective reaction to reported information security events and weaknesses. The procedures should ensure that all reported events are reviewed and investigated where appropriate, that recovery procedures are triggered and that roles of suitable seniority are involved in reviews. ISO/IEC 27002, 5.24 provides a list of measures that should be applied to properly manage information security incidents.

Auditing guidance

The auditor should check that information security incident management procedures are in place, and that they are compatible with likely reporting scenarios, e.g. as described in ISO/IEC 27002, 6.8. They should also check that all reported information security events and weaknesses are reacted to appropriately. ISO/IEC 27002, 5.24 describes the procedures that should be in place to handle and recover from system failures, errors, security breaches, etc., including contingency arrangements and auditing activities.

5.25 Assessment and decision on information security events (ISO/IEC 27001, A.5.25)

"The organization shall assess information security events and decide if they are to be categorized as information security incidents."

Implementation guidance

Each event that is reported should be reviewed by the information security contact to whom it was reported, to ascertain its level of impact. This should be used to determine whether the event should be categorised as an information security incident, and the decision and supporting justification recorded in the incident reporting system. The point of contact may not be the final arbiter of classification. If the organisation has a team that handles information security incidents, this team should confirm the initial classification.

This is the initial triage phase of information security incident management, which ensures that the efforts of incident response teams are focused upon the correct events, and that serious incidents are handled promptly. To ensure that decisions are consistent, the organisation should have defined incident categories (e.g. critical incident, minor incident, etc.), with appropriate supporting guidance. The categories should be periodically reviewed for clarity, relevance and usefulness, and updated as required. The individuals responsible for triage should have a clear understanding of the organisation's requirements for timely triage, based on its legal and contractual requirements to notify customers/other entities of a breach.

Auditing guidance

The auditor should ask for records of incidents and timeline documentation, and check that there is a point where a clear determination was made of whether an event should be classified as an incident. They should ask how this was determined and verify that there is a consistent, repeatable

process in place. Guidance documentation explaining how to
categorise an incident should be readily available to all
points of contact, and a list of points of contact available to
all staff. The auditor should interview points of contact to
verify whether they understand the incident categorisation
and reporting process. There should also be a process for
notifying points of contact should the guidance change.
There should be targets that are consistent with the
organisation's stated timescales for notification.

5.26 Response to information security incidents (ISO/IEC 27001, A.5.26)

*"Information security incidents shall be responded to in
accordance with the documented procedures."*

Implementation guidance

Information security incidents should be responded to in
accordance with management requirements, and as
documented (see 5.24). ISO/IEC 27002, 5.26 provides a list
of actions that should be carried out to properly manage
information security incidents. The organisation should
nominate a point (or points) of contact to ensure that a clear
reporting line is used.

One particularly important consideration regarding incident
response is that the organisation must decide very early on
whether evidence will need to be collected (see 5.28). This
will shape the whole incident response process.

Auditing guidance

The auditor should confirm that relevant activities, such as described in ISO/IEC 27002, 5.26, are properly documented in procedures. The procedures should identify all responsibilities. Records of incidents should demonstrate that appropriate management control is exercised, and the auditor should confirm that all information security incidents and their follow-up activities are properly recorded.

5.27 Learning from information security incidents (ISO/IEC 27001, A.5.27)

> *"Knowledge gained from information security incidents shall be used to strengthen and improve the information security controls."*

Implementation guidance

In addition to detecting and taking action to resolve information security incidents, it is important that the organisation (and the relevant roles within the organisation) learns from these incidents to avoid future occurrences, or if they cannot (or should not) be avoided, to ensure they can be dealt with more effectively. This is also part of the Performance evaluation and Improvement aspects of the ISMS (see Clauses 9 and 10 of ISO/IEC 27001), as the evaluation of incidents helps identify where controls did not work as intended, and where improvements are necessary. Learning from information security incidents will provide useful information about actions that need to be taken to enhance security, and suitably anonymised case studies should also be used judiciously in training and awareness programmes.

Auditing guidance

The auditor should review examples of how the organisation has responded to information security incidents, and software and system weaknesses, in the past, and verify that there is a consistent approach to ensure that learning from incidents is captured and promptly applied. The process to learn from incidents should include the implementation of additional controls or procedures to avoid reoccurrences, limit the damage, collect evidence, or allow a quicker and more efficient reaction in the future. For example, anonymised incidents should be used in training and awareness programmes to give real-life examples.

Where it appears that an insufficient number of incidents have occurred, or that information or evidence is not available to learn from, the auditor should check whether the reporting procedures for information security events and weaknesses are being correctly used, and are designed to support improvement work.

5.28 Collection of evidence (ISO/IEC 27001, A.5.28)

> *"The organization shall establish and implement procedures for the identification, collection, acquisition and preservation of evidence related to information security events."*

Implementation guidance

It is important that an organisation ensures the collection of admissible and complete evidence for any information security incident that might result in formal legal proceedings. The organisation should have guidelines and procedures for the collection of evidence that ensure appropriate admissibility and completeness of the evidence.

Where evidence is collected for legal purposes, it should be managed and stored to maintain the chain of custody and guarantee that no one can modify or destroy it without authorisation. The organisation should also ensure that the evidence is available in a timely manner and in a form that is required by a court. ISO/IEC 27002, 5.28 describes the factors that should be taken into account when writing procedures for the collection of evidence.

If the organisation is carrying out any investigative or forensic work, this should only be done using forensically sound copies of any evidence that might be required later, to ensure it can be proved that the evidence has not been altered or tampered with.

Auditing guidance

Collection of evidence is important to be able to provide adequate support in legal procedures and actions that might take place as a result of information security incidents, such as a breach of civil or criminal law. The auditor should confirm that the organisation has procedures in place to collect evidence. Auditors should check that these procedures include appropriate considerations, for example those in ISO/IEC 27002, 5.28, and that they ensure that:

- the information collected conforms to applicable standards or codes of practice for such evidence to be deemed admissible as evidence; and
- the quality and completeness of such evidence is appropriate.

Auditors should check where collected evidence is stored, and whether it is possible for unauthorised persons to access, modify or destroy such evidence. In addition, auditors should consider the conditions when the evidence collection process needs to be activated. Collection of evidence should start at an early stage to ensure that no evidence is destroyed or contaminated.

5.29 Information security during disruption (ISO/IEC 27001, A.5.29)

> *"The organization shall plan how to maintain information security at an appropriate level during disruption."*

Implementation guidance

In the case of any serious unexpected event, especially one that affects business continuity, it is important that information security does not 'fall by the wayside'. If a level of security is necessary when things are going well, then it is also necessary when things are going wrong – unless the organisation wishes to escalate a business continuity event into a combined business continuity event and information security incident.

It is therefore essential that information security is considered and included in the overall business continuity management process. The process that the organisation uses to manage and recover from crises (whether it is called the disaster recovery process, the business continuity process or something else) should have integrated into it the principle that information security remains important in a crisis, and should make this possible. The organisation should identify and document its information security requirements for business continuity. These should be based on clearly realised and authorised objectives for information security in atypical circumstances. The organisation should also ensure that the levels of information security provided in suppliers' business continuity arrangements are sufficient for its requirements.

Both information security specialists and business continuity/disaster recovery specialists should be involved in integrating information security continuity into the plans.

Auditing guidance

The auditor should confirm that the organisation has a clear statement of its objectives for information security continuity. To support this, the auditor should check that the organisation has documentation that pertains to the management of serious adverse events (by whatever name). There should be a policy statement regarding preservation of information security during the management of, and recovery from, these events. There should be evidence of a risk assessment, leading to a list of requirements for information security during an adverse event.

The plan for handling adverse events should include information security considerations throughout the process, including within the business impact analysis stage. Auditors should confirm that the scope and details of this plan fulfil the organisation's information security requirements, and that it has been signed off by management.

Staff responsible for managing crises should have access to (and understand) instructions on how information security should be managed. Staff with information security responsibilities should have access to (and understand) instructions on their responsibilities during a crisis.

In addition, the auditor should check that:

- all responsibilities are agreed and assigned;
- all procedures defined in the plans are documented and implemented according to the implementation schedule; and
- all staff are aware of and understand what they are supposed to do in case of emergencies and business interruptions.

Testing of continuity plans is essential, and auditors should determine the testing schedule, which may be defined in the planning framework or in the plan(s) itself. It is unlikely that plans with any degree of complexity will work perfectly first time. Individuals need to follow the procedures to handle the situation effectively, and this approach will only work if it has been practised.

5.30 ICT readiness for business continuity (ISO/IEC 27001, A.5.30)

> *"ICT readiness shall be planned, implemented, maintained and tested based on business continuity objectives and ICT continuity requirements."*

Implementation guidance

This control speaks to the need for availability of information, obviously – in cases where business disruption takes place, is there a good plan to ensure that, for instance, services undergo graceful degradation in the face of increasing load? A good illustration of this is a website – as load on the system increases, the service can be designed such that the most important content is prioritised (for example, by being transmitted first) and cosmetic aspects such as images are deprioritised. This ensures that the core functions of the website are the last to fail under load.

There are a plethora of excellent processes and standards in the field of business continuity that include ICT continuity in their scope, including ISO/IEC 27031, ISO 22301, ISO 22313 and ISO/TS 22317. The organisation should use these to design and implement structured processes (e.g. a business impact analysis) supported by prioritising functional aspects of the ICT environment and relating them to the operations of the organisation in a meaningful and actionable way.

The relevant stakeholders should be involved from the start to enable a coherent approach to prioritisation, and to determine how long is too long – can the service under review withstand long periods of unavailability? This feeds into the RTO/RPO, etc. specifications, which are then useful in communicating organisational expectations to the IT functions.

If the organisation is dependent on the availability and reliability of services provided by a supplier, it should also check the plans that the supplier has put in place to deal with interruptions of its services. It might also be worthwhile testing these situations to know what to expect in an emergency situation.

NB: The focus here is exclusively on managing availability loss and restoration, without the consideration of appropriate levels of confidentiality or integrity – yet without these attributes, the service that is restored will not be fit for purpose. So it would be concerning to see this control implemented without the preceding control (5.29) also being selected.

Auditing guidance

The auditor should review other standards in the business continuity space that the organisation may be certified to (as long as the scope of the certification is appropriate). They should interview key stakeholders (internal and external, if relevant and practical) regarding priority environments and services, expected recovery times, and the order in which they would expect to see service restored to business-as-usual levels. These should be adequately and consistently documented, and communicated via a consistent route to all parties required to be aware of the information.

The auditor should then follow through with the IT function, interviewing staff accountable for service continuity and restoration to see if they have the same understanding of priorities. The recovery processes should be reviewed to see if they can achieve the targets – and, indeed, to see if the specified targets are meaningful in the operational context. Plans or playbooks should be readily available to the parties who are intended to participate, and should be clear enough to be used under situations of high stress.

Records of ICT continuity testing should be reviewed to ensure that the most important systems and probable failure modes have been tested, and that findings have been used to improve effectiveness, as usual (also see Clause 10.1 of ISO/IEC 27001). Records of ICT continuity events, if any exist, should be reviewed in a similar manner.

5.31 Legal, statutory, regulatory and contractual requirements (ISO/IEC 27001, A.5.31)

> *"Legal, statutory, regulatory and contractual requirements relevant to information security and the organization's approach to meet these requirements shall be identified, documented and kept up to date."*

Implementation guidance

The organisation should identify and document all statutory, regulatory and contractual requirements to ensure their fulfilment. It should also identify the approach that it will take to coordinate and meet these requirements. Especially when thinking of conducting business in other countries, the identification of applicable legislation should be supported by an expert, e.g. a lawyer. Special attention is also required when conducting online business or trading to ensure compliance with all relevant legislation in the countries involved. As the statutory, regulatory and contractual requirements will change over time, it is important that the organisation has appropriate change control procedures in place that incorporate these considerations.

The legal and regulatory requirements and rules for the use of cryptographic controls, and the effort and resources necessary to comply with them, should be assessed with particular care, as cryptographic systems may be subject to special or different rules depending on the jurisdiction. The results of these assessments should be taken into account in the decision about the use of cryptographic controls. This assessment should not only include the laws and regulations applicable for encryption controls, but also the legal environment for the use of digital signatures and other electronic communications. ISO/IEC 27002, 5.31 lists items that should be taken into account when creating policies and procedures.

Auditing guidance

The auditor should inspect the actions that have been taken to identify, document and comply with all applicable statutory, regulatory and contractual requirements. They should check that no applicable legislation, regulations or contracts have been forgotten or missed by mistake. This may require specialist expertise.

The auditor should confirm that the organisation has controls in place to comply with the requirements it has identified. Responsibilities for these controls should be identified and documented, and those responsible should be aware of their responsibilities. For example, someone should be responsible for keeping the identified statutory, regulatory and contractual requirements up to date, as these will change over time. Changes to requirements should be traceable to changes to implemented controls.

The organisation should present to the auditors the actions it has taken to identify applicable legislation and regulations for cryptographic controls, and the legal advice it has taken where necessary to ensure compliance. The controls that are taken to fulfil these requirements should be documented, implemented and maintained. The auditor should check that the implementation of the policy for the use of cryptographic controls as described in 5.31 is commensurate with the legal requirements identified. The auditor should also find out how the organisation is tracking changes to legislation and regulations in the jurisdictions within which it operates.

5.32 Intellectual property rights (ISO/IEC 27001, A.5.32)

> *"The organization shall implement appropriate procedures to protect intellectual property rights."*

Implementation guidance

Organisations are vulnerable to failure to comply with restrictions on usage of copyright material. There is a serious risk of legal action being taken against the organisation and individual staff where, for example, software is being used on more than the number of systems it is licensed for.

The organisation should put rules in place for the handling of material, and these rules should take into account all types of restrictions, for example software or document copyright, design rights, trademarks, patents and source code licences. Staff should be made aware of the rules. For software, it is especially important to make staff aware of these rules, and inventory checks should be carried out at least annually to provide assurance that all software in use (that is, software loaded on the system) is properly licensed. Documentary records should be maintained of the inventory of software on each system (see also 5.9).

With regard to AI-generated material, it is also particularly important to establish who holds the copyright of the material generated; for example, was the original source material taken from sources on the Internet that are themselves copyrighted? And does the AI service provider claim ownership of all the information that the AI produces, even if it is using information provided by the end user?

It is also important to note that rules relating to copyright, etc. vary significantly in different countries, so the organisation should take this into account when operating internationally.

ISO/IEC 27002, 5.31 provides guidelines on the protection of copyright material. The organisation and all staff should be aware that copyright infringement can lead to legal action that may involve criminal proceedings.

Auditing guidance

Auditors should confirm that the organisation has procedures in place to protect the intellectual property rights of copyrighted information and software. These procedures should describe rules for handling material that is marked as copyright, design rights or trademarks, and employees should be aware of how to handle such material. These rules should address the handling of all copyright material, irrespective of its form. The auditor should confirm that users are aware that any unauthorised use or copying of intellectual property rights material or software might lead to legal action.

There should be strict controls on the use of software and other copyrighted material (for example subscription-based online content) in the organisation. Auditors should investigate what licences have been purchased and how compliance is maintained. Many commercial packages provide licence agreements on the packaging, which come in various forms – there is no common format for information such as number of users, restrictions to use, etc. Further information must be gathered to check that each software package is being used in compliance with the related licensing agreement.

One way to address these issues is to draw up a table of resources protected by copyright, and then identify the key aspects of each licence, together with a record of use. The auditor should look at the use of development tools and libraries. Have these been used correctly? With bespoke software developed for the organisation, look at the development or support contract. Is access to source code provided? Can in-house changes be applied? Are there restrictions on the use or location of the software? Ensure that the responsible personnel in the organisation are fully aware of their obligations regarding software copyright. The auditor should also check a random sample of computers (both workstation and server) for unlicensed material.

5.33   Protection of records (ISO/IEC 27001, A.5.33)

"Records shall be protected from loss, destruction, falsification, unauthorized access and unauthorized release."

Implementation guidance

Organisations will have various essential documents and records, such as accounting records, database records, transaction logs, audit logs and operational procedures, that need to be retained and should be protected from loss, breach of confidentiality or modification. These items should be listed in the asset inventory (see also 5.9), and appropriate controls selected and implemented to ensure the protection of these records until the end of their retention period. The continued presence of the items should be confirmed by a documented inventory check at least annually.

For example, under various regulations, organisations are required to maintain business records of certain types for periods of up to ten years, and an organisation is open to prosecution where this has not been carried out.

When keeping records for such a long time, due consideration should be given to deterioration of the media on which these records are stored, and the organisation should ensure that tools (e.g. microfiche readers, software and cryptographic keys) are still available, and it is still possible to access records up to the end of their retention period. ISO/IEC 27002, 5.33 provides further guidance on the protection of organisational records.

Auditing guidance

The auditor should confirm that all records required by the organisation, including for legal or regulatory purposes (e.g. financial records, customs records, legal records and environmental records), are identified, and that all retention requirements are complied with. Legal and regulatory requirements will vary from country to country, and the organisation needs to be aware of, and comply with, all applicable requirements. The auditor should check that this has been done and verified by the appropriate personnel. The storage arrangements (including security) and the requirements for review and disposal should all be defined in procedures.

There should be an inventory of records, and auditors should check this for accuracy. Some documentation might now be held electronically, either because that was the original format or because they have been scanned. The auditor should check that the organisation has reviewed the legal admissibility of this storage medium and is complying with any additional requirements for preservation of the integrity or availability of these records. The auditor should also check that the organisation has considered the risk of media deterioration and the resulting loss of accessibility of data if electronic storage media are chosen, and that it has measures in place to ensure the availability of the necessary cryptographic keys, if encryption was chosen to protect the records (see also 8.24).

5.34 Privacy and protection of personal identifiable information (PII) (ISO/IEC 27001, A.5.34)

> *"The organization shall identify and meet the requirements regarding the preservation of privacy and protection of PII according to applicable laws and regulations and contractual requirements."*

Implementation guidance

In many countries, legislation or regulation is in place to protect the privacy of personally identifiable information (PII). Failure to comply with such legislation can leave the organisation open to prosecution and a fine, or at least to serious loss of image and reputation if it became public.

Several laws also specify a number of requirements for the collection, processing, accessibility and protection of PII in any form. Failure here can also lead to prosecution. Finally, prompt breach notification is a significant requirement in some legislation. In addition to this, clients or other entities may have provided PII to the organisation for business purposes, and may have imposed additional requirements for its protection, over and above local laws. This situation may occur if the PII is held in a jurisdiction where additional protection is not required by law.

If the organisation stores, processes or transmits any PII, and if there is applicable legislation or regulations, or contractual terms, it should develop and implement a policy that ensures that no requirements are disregarded. An inventory of PII should be kept, including metadata such as the purposes for which each set of PII is being used, and the justification for its retention. A senior role should be given accountability for compliance. Responsibility for handling this information should be assigned in a manner that complies with legal and contractual requirements, and staff trained regularly. Procedures should be designed to ensure that changes in use are managed and approved, and recorded in the asset inventory. A documented review of PII holdings should be carried out at least annually, and compliance with all requirements stated in the relevant laws or regulations should be verified. It may be helpful to maintain a dedicated matrix that relates all PII protections clearly back to the source(s) of their requirements, so that it is easier to update protections when laws change, and to demonstrate compliance for all purposes.

Auditing guidance

Careful control of PII is necessary to comply with the applicable legislation and regulations that might apply. Many countries, for example in Europe, have well-developed data protection legislation. The legislation might also require the organisation to register its use of PII.

The auditor should confirm that the organisation knows where the PII it handles is located, has identified all relevant legislative or regulative requirements, and has put policies, procedures and controls in place to comply with them. The auditor should also look at the type of data held. Is it necessary? Has it been validated? Is it transmitted or otherwise conveyed outside of the organisation? Who has access to this data, and is it necessary for their job function? A sample of roles handling PII should be interviewed to ensure that they have had additional training and are clearly aware of their responsibilities and of any specific controls that are to be applied to PII.

The auditor should check that the organisation monitors changes to requirements in this area – new, tighter restrictions can be introduced with specified periods for compliance. Is there sufficient awareness within the organisation? Are there plans to introduce compliance within the time frame? Auditors should ensure that they themselves are fully up to date with this area of legislation.

5.35 Independent review of information security (ISO/IEC 27001, A.5.35)

"The organization's approach to managing information security and its implementation including people, processes and technologies shall be reviewed

> *independently at planned intervals, or when significant changes occur."*

Implementation guidance

As with all business activities, the organisation's approach to information security and its implementation should be reviewed from time to time to ensure that everything is still suitable and effective. Additional reviews should be carried out when major changes are planned, or major unplanned changes have occurred. These additional reviews should be initiated by a clear trigger that is linked to the organisation's change and incident management processes.

The results of these reviews should be reported to management. Each review should be carried out by an independent body (either within the organisation or outside) to provide assurance to senior management that the organisation's ISMS practices are adequate and effective.

'Independent' does not exclude an internal review, provided that the reviewer has appropriate hierarchical independence from the environment/scope being reviewed. An internal audit department would be appropriate. However, a small organisation might find it necessary to use an external party for the review. A certification audit undertaken by a suitably accredited organisation would also satisfy the requirements of this control.

All results of independent reviews should be recorded, as well as any corrective action that is taken if the independent review identifies areas for improvement.

Auditing guidance

It is important for the auditor to check that independent reviews of the organisation's approach to information security and its implementation are taking place, and that they are carried out by an independent party. Without this, objectivity cannot be achieved. A third-party audit can satisfy this requirement. In cases where third-party audits are not being performed, the requirement for independence can be satisfied via internal auditors, management, or other bodies external to the security practitioners/implementer teams.

Check the audit schedule, but also the process for identifying the need for a supplementary audit when a major change is planned. What are the criteria in use? Ask for examples showing where this has occurred.

The auditor should check the records of the independent reviews, and should verify that identified corrective actions have been implemented as planned and have had the desired effect. The results of other reviews, such as policy reviews as described in 5.1, should also be taken into account.

5.36 Compliance with policies, rules and standards for information security (ISO/IEC 27001, A.5.36)

> *"Compliance with the organization's information security policy, topic-specific policies, rules and standards shall be regularly reviewed."*

Implementation guidance

If the organisation is expending effort and resources in implementing controls, it should ensure that these controls are working effectively, as, for example, required by ISO/IEC 27001.

Managers should review compliance in their area of responsibility with security policies, controls and standards. This can take place through a formal review, and/or through spot checks that can occur at any time during normal work. A combination of both techniques is possibly most successful. The complexity of information systems – for example, servers, networks and firewalls – means that despite best intentions they might still be in an insecure state. A full technical review should therefore be carried out at suitable intervals, determined by risk assessment, to detect any technical nonconformities.

Operational information systems require skilled analysis, aided sometimes by tools that automate certain types of tests. Indeed, the use of such programs allows some tests to be carried out more frequently, as well as more swiftly, and with a lower chance of manual input errors.

These checks should only be carried out by, or under the close supervision of, competent, authorised persons. The integrity and availability of systems under test could be jeopardised should an insufficiently skilled person attempt this work. In addition, certain types of compliance tests, such as penetration testing, can result in criminal charges (related to computer misuse) if accidentally directed at the wrong systems. Also note that some testing tools may be identified as attack tools, since their purpose may be to try to compromise a system, and their use by unauthorised users should be treated as an information security incident. Access controls should generally prevent unauthorised persons from carrying out technical compliance reviews (see 5.15). The installation of tools used for testing should be strictly controlled (also see 8.18 and 8.19).

Reviews should be planned and documented. Results, nonconformities, actions and follow-up activities should be recorded and traceable to each other, so that, for example, a nonconformity can be traced to the action that was initiated to correct it, and there is also a record of the impact of that action. Clause 10.1 of ISO/IEC 27001 requires nonconformities to be identified and rectified. It also requires the organisation to look out for the chances of similar issues in other areas.

Managers should react to identified non-compliances as described in ISO/IEC 27002, 5.35. A documented record of each review should be maintained, noting non-compliances, agreed action and follow-up. The managers should also report the results of their reviews into the internal review process.

Auditing guidance

To determine the degree to which security policies and procedures are being complied with, the auditor should confirm that management has procedures in place to regularly review this compliance appropriate to their area of responsibility. These reviews should be fully documented, followed up to ensure resolution of non-compliant items and reported on to the internal review (see also 5.35), and as required to senior management.

The auditor should confirm that the organisation also has scheduled technical checks in place to ensure that its information systems conform to its security implementation standards. There should be a plan for technical conformity checking, showing what needs to be covered, and the frequency and methods employed. It is important that this type of conformity checking is performed by suitably competent and authorised personnel.

Records of checks should be traceable to findings and remedial actions, and finally to testing to ensure that the remedial actions had the desired effect. Ask if the organisation looks for patterns in issues and how it identifies and follows them up. A technical issue found on one system could exist on many more.

If a tool is used, the auditor should check what aspects of the information systems are being reviewed – it could purely be monitoring or conducting an audit of facilities. Has it been validated in any way? Check the individuals completing or reviewing the reviews, as compliance can only be effectively assessed by technically competent personnel. Personnel carrying out reviews should also have been authorised to do so.

Managers should also be able to provide procedures that they use to respond to non-compliance. These procedures should ensure that the root cause(s) for the non-compliance are identified, that any necessary actions are taken to avoid recurrence of the non-compliance, and that the appropriate corrective actions are identified and successfully implemented. The auditor should ask to see records showing how these procedures have been used, and follow action entries through to practical evidence of their implementation (e.g. a system configuration, record of remedial training, or updated supplier contract).

5.37 Documented operating procedures (ISO/IEC 27001, A.5.37)

> *"Operating procedures for information processing facilities shall be documented and made available to personnel who need them."*

Implementation guidance

As with all the controls in this standard, the scale of procedures should be appropriate to the size and complexity of the organisation. A large organisation with many staff might require more comprehensive and detailed procedures than a small organisation, where a few thoroughly experienced staff cover the whole operation. The documentation requirements for these procedures might also vary. In any case, the organisation should ensure that sufficient documentation is available to address typical activities in the day-to-day working environment, e.g. start-up and shutdown procedures, backup, equipment maintenance, mobile working, media handling, CCTV usage, mail handling, and safety. Typical instructions that operating procedures should include are described in ISO/IEC 27002, 5.37.

Inadequate or incorrectly documented procedures can result in system or application failures, causing loss of availability, failure of data integrity and breach of confidentiality. Complicated or infrequently used procedures provide opportunities for mistakes and should be avoided where possible. Operating procedures should be treated as formal documents, changes to which may only be approved by authorised persons.

Many organisations outsource the management of their computers and communications to a specialist facilities management organisation. One way of ensuring that appropriate security is in place is to use sufficiently detailed contracts, and to check whether the other organisation is ISO/IEC 27001 compliant (see 5.22 for more about working with service providers).

Auditing guidance

Auditors should check the organisation's operating procedures and confirm that these are appropriately documented and are being applied throughout the relevant parts of the organisation. To be able to check procedures for completeness, auditors should have a general understanding of the operational processes and workings of the organisation.

In addition, the curation of procedure documents should be checked. A check should be made to ensure that it is not possible to modify procedure documentation without appropriate authorisation, that proper version control is in place and that the latest version is readily accessible to all those who need it.

Another aspect to check is the level of compliance with procedures. Is it possible to circumvent these procedures or any associated controls? Are they commonly circumvented? Are the employees aware of these procedures, and do they know which procedure to use and where to find it, if needed? Are they using the latest version?

Responsibility for network service, or cloud service, operation and administration is often held by a separate department, or even a separate organisation (also see 5.21).

The auditor therefore needs to understand the arrangements in place, and confirm that the necessary levels of service and procedures are properly documented. In some areas, detailed work instructions will be needed. There is likely to be considerable use made of supplier documentation, so this should also be checked for relevance and availability. This issue is addressed in more detail in 5.22.

CHAPTER 6: PEOPLE CONTROLS (ISO/IEC 27001, A.6)

6.1 Screening (ISO/IEC 27001, A.6.1)

> *"Background verification checks on all candidates to become personnel shall be carried out prior to joining the organization and on an ongoing basis taking into consideration applicable laws, regulations and ethics and be proportional to the business requirements, the classification of the information to be accessed and the perceived risks."*

Implementation guidance

Screening is an essential control that can prevent the organisation employing the wrong person. Identification checks, CV reviews, checks of qualifications and verification of character references are possible elements within the screening process, but legal constraints may specify the type and depth of checks permitted. Whatever screening and data collection takes place, care should be taken to ensure that all applicable legislation and regulations are complied with.

Where the proposed position provides access to sensitive, critical and/or personally identifiable information, it is highly advisable to verify the nature of a candidate's responsibilities in previous similar positions. However, some organisations will not, as a matter of policy, offer any detail or opinion other than confirmation of the period employed and the previous position held. Gaps or apparent irregularities in employment should be questioned.

All exchanges and interviews should be fully documented and retained on file throughout employment and for a reasonable period after it ceases, or after rejection of an application pending any possible appeal by the applicant. The required screening processes should take place for all people working within the scope of the ISMS, irrespective of whether these are employees, contractors or third-party users.

Auditing guidance

The auditor should collect evidence that screening procedures for personnel recruitment (including contractors, third-party users and temporary staff) are being consistently applied and include appropriate verification checks. ISO/IEC 27002, 6.1 lists the items to be covered. Organisations should not rely on employee-supplied CVs, endorsement letters or qualifications without suitable independent verification of the data supplied. The auditor needs to check any follow-up actions, such as conversations with referees, which should be documented. Managers and recruitment staff should be interviewed to establish that they are aware of their responsibilities for evaluating and reviewing the background checks for staff in their area of responsibility. The auditor should also check that all information related to personnel verification checks is handled in accordance with all relevant regulations and legislation (e.g. data protection; see 5.34).

6.2 Terms and conditions of employment (ISO/IEC 27001, A.6.2)

> *"The employment contractual agreements shall state the personnel's and the organization's responsibilities for information security."*

Implementation guidance

It is important that employees, contractors and third-party users are aware of their security and legal responsibilities regarding the handling of information, the classifications and use of information processing facilities, and the consequences of not complying with security or legal requirements. This also extends to any contractual obligations that the organisation has entered into that might affect the employee's, contractor's or third-party user's scope of work. Any such responsibilities should be included in terms and conditions of employment.

It is also important that employees, contractors and third-party users sign a confidentiality agreement (see also 6.6) before starting work, and that they understand that such responsibilities may extend beyond their normal working environment and working hours, as well as home working, working on customers' sites and any other form of remote working. Some confidentiality agreements may persist beyond the termination of the individual's employment with the organisation, if legally permitted (see 6.5).

In addition, the organisation's responsibilities for handling the personal data of employees, contractors and third-party users should be stated, for example compliance with data protection legislation (see also 5.34).

Auditing guidance

Auditors should check whether the terms and conditions of employment accurately describe both the employer's and the employee's, contractor's or third-party user's responsibilities for information security. These descriptions should cover all security-relevant aspects of the employee's job, including responsibilities applicable to legal requirements, responsibilities related to classified information, working outside the organisation or outside normal working hours, and those responsibilities that might extend beyond the employee's contract. The terms and conditions should also describe the actions that will be taken if employees do not fulfil their security responsibilities.

The auditor should check that agreement to, and signing of, the terms and conditions of employment is a requirement before any work starts. Employees, contractors and third-party users should also be required to sign a confidentiality agreement (see also 6.6) before accessing any confidential information. The auditor should confirm that procedures are in place to ensure that the terms and conditions of employment are updated if the employee's security responsibilities change in any way, e.g. taking on new roles or using new or different information processing facilities.

The auditor should also check that the organisation's responsibilities for handling personal data are clearly stated, e.g. compliance with data protection legislation (see also 5.34).

6.3 Information security awareness, education and training (ISO/IEC 27001, A.6.3)

> *"Personnel of the organization and relevant interested parties shall receive appropriate information security awareness, education and training and regular updates of the organization's information security policy, topic-specific policies and procedures, as relevant for their job function."*

Implementation guidance

The organisation is vulnerable to the activities of untrained employees, contractors and third-party users. There is a risk of them producing incorrect and corrupted information or losing it completely. Untrained personnel can take wrong actions and make mistakes through ignorance.

All personnel should be trained in the relevant policies and procedures, including security requirements and other business controls. They should also be trained to use all the IT products and packages required of their position, as well as in the relevant security procedures. The organisation should consider when training should be repeated and updated.

Training might be required at different levels as follows.

a) Basic security awareness: every employee, and where relevant, contractor and third-party user, should be given a foundational level of security awareness training. The foundation training should explain the organisation's security policy, objectives and procedures that everyone is required to follow.

b) Supplementary security training: staff with special responsibilities for security (not only security-dedicated roles) should, in addition to basic training, be provided with relevant, specialist training. A training plan should be developed for these individuals, taking into account both the specific knowledge and skill required for their role, and their existing capabilities, understanding and needs.

The general development of security knowledge can benefit significantly from employees attending suitable conferences and carefully selected events, which are frequently free. Ensure that training suppliers use appropriately qualified staff, and that the syllabus is clear and consistent with the organisation's requirements. Training that reflects the ethos and culture of an organisation will be remembered better and more usable than generic training.

The following approach will ensure that training and awareness have the best chance of effectiveness.

- **Keep it current:** Regularly update security education and awareness programmes to address emerging threats and ensure alignment with the latest organisational policies and procedures.
- **Availability:** Make the material supporting training (including procedures and policies) readily available to employees.

- **Reinforcement:** Refresh awareness as necessary via refresher courses or tests to determine whether additional training is required.

- **Inclusive:** Ensure that training caters to all levels of the organisation, including contractors and third-party users, reflecting the need for an holistic approach.

- **Engagement and understanding:** Use engaging training methods that foster a deep understanding of security policies and encourage active participation in security practices.

- **Performance metrics:** Establish training records and use clear metrics to assess the effectiveness of the training programmes. This could involve regular testing, feedback sessions, and practical exercises that simulate security incidents (see also 5.24 and Clause 9.1 of ISO/IEC 27001). All training, test results and relevant event attendance should be recorded in the individual's training record.

- **Integration with HR processes:** Integrate a review of necessary security training, and supplementary training as needed, into HR management processes such as onboarding, role changes and offboarding to maintain security awareness throughout the employee lifecycle (see also 6.5).

Auditing guidance

This control is applicable to all employees, including users of information processing facilities such as system administrators, managers and application users, as well as senior management and those processing any form of information (e.g. paper based, telephone, etc.). It is also applicable to contractors and third-party users, and anyone else with access to information or services within the ISMS.

The first point to check is the appropriateness of the training. This should be consistent with the job and the related security responsibilities. How is it provided: internally or externally? If internal, is it a formal course or general 'on-the-job'-type training? Who is providing the training, and are they suitably qualified? If the training is informal, is there some definition of what has been covered? If the training is external, who has approved the supplier? Is the supplier an accredited or industry-recognised provider of training? What records exist and do they reflect the nature and depth of training given?

At a minimum, organisations should have some form of induction training that is given to all employees, contractors and third-party users. This should cover the general principles of security, the information security policy, areas of applicability, etc. This should be formally recorded in individual records. In addition, the auditor should ensure that sufficient training for those with more complex security responsibilities is in place, all training material is up to date and the training is provided in time for the job to be carried out. Check the records for different types of job function to ensure that sufficient training is provided, and that it is provided before access to information or services is given.

The auditor needs verification that staff are aware of the content of the information security policy, how they contribute to the effectiveness of the ISMS and the implications of not conforming with the ISMS's requirements (ISO/IEC 27002, 7.3). This can be obtained by interviewing staff members whom the auditor randomly selects.

Training should be being repeated as appropriate to ensure that it is current and reinforced, and when the information that people should know, or the skills they need, change. On a wider scale, the organisation should be able to show that it regularly reviews the effectiveness of its training programmes and makes adjustments based on feedback and changing security landscapes. For example, records of incidents could be used to determine topics for particular focus in the next quarter's training programme.

There will be situations, particularly with technical aspects, where experience or previously acquired qualifications are claimed in lieu of formal training. Auditors need to take a pragmatic approach and view the effective result of formal training, qualifications and experience when looking at the skills of individuals and how they fit with their roles. If previously acquired experience is claimed, make sure it is current and relevant. Also check what environment the experience was gained in, and whether it has been verified in a suitable fashion. Many organisations rely too heavily on what individuals claim in CVs – an inadequately trained or inexperienced individual in a key position can cause major damage to vital assets, so it is important that the organisation treats training verification seriously. This also relates to the checks that should be made on recruitment (see 6.1).

6.4 Disciplinary process (ISO/IEC 27001, A.6.4)

> *"A disciplinary process shall be formalized and communicated to take actions against personnel and other relevant interested parties who have committed an information security policy violation."*

Implementation guidance

This might be a sensitive issue in organisations, but it is important that a process is in place to react properly to, deter and correct non-compliance. This prevents a decline in standards and an increase in insecurity.

> *NOTE: it is necessary for policy to address controls compliance, so a non-compliance with security controls will also be a breach of policy.*

The disciplinary process should be shaped by the organisation's culture and management practices. It should be documented, and people within its scope, including third parties where applicable, should be aware of how it works. The disciplinary process should ensure consistent and fair treatment, and should only be initiated after there is sufficient evidence that a non-compliance with policies has taken place. If the disciplinary process is not implemented correctly, the organisation may be liable to potential claims of unfair dismissal or other personal infringements.

The process that is used should follow that which is documented. Different levels of non-compliance should be considered in the disciplinary process, so that the response is proportionate. Compliance incentives (which can be included in training programmes) can act as an excellent foil to sanctions for non-compliance.

Auditing guidance

Auditors should check that all employees are aware of the disciplinary process, that it is initiated following an appropriate trigger, and that it provides fair and consistent treatment to all involved.

Auditors should also check recorded security incidents, look at the criteria for disciplinary action and verify that it has been initiated when criteria have been met, and check that follow-up action has always been carried out to completion of the matter. Auditors should also verify that procedures are in place to ensure that the disciplinary process is only used if there is sufficient evidence that a security breach or non-compliance has occurred.

6.5 Responsibilities after termination or change of employment (ISO/IEC 27001, A.6.5)

> *"Information security responsibilities and duties that remain valid after termination or change of employment shall be defined, enforced and communicated to relevant personnel and other interested parties."*

Implementation guidance

Many problems can occur if termination or change of employment is not handled appropriately (see also 5.11 and 5.18). This could include excess permissions being retained by an individual who changes roles, resulting in a loss of segregation of duties. This could also include an individual retaining access to sensitive staff records after they have left the organisation, which breaches data protection laws.

There should be processes in place to ensure that all rights for logical and physical access are removed as and when the job function terminates. A specific role within the organisation should be responsible for this, including ensuring that the necessary communications between departments take place in case of termination or change of employment. On termination, all equipment and information belonging to the organisation, or to its customers, should be returned as per company policy.

The responsibilities for termination or change of employment should also include any ongoing security or legal requirements that need to persist after any business relationship has ended. The most typical example is the confidentiality agreement. All such responsibilities and duties should be covered in the employee's, contractor's or third-party user's contract.

An important point that is sometimes overlooked is that these controls should be applied not only in the case of employment termination, but also when an employee, contractor or third-party user changes role. A change of role should be handled as a termination of one role and the beginning of another. All logical and physical access rights related to the old role should be rescinded, and all assets related to the old role should be returned.

Auditing guidance

The auditor should confirm that the organisation has procedures in place for termination and change of employment. The auditor should look for records showing that these procedures and responsibilities are clearly assigned and executed, and that responsibilities have been appointed for the termination of logical and physical access rights and the return of assets (see also 5.11). There should be procedures in place to ensure that all parties involved in these actions are notified if a termination or change of employment takes place, as well as other parties who need to be aware of the event. Records of these notifications should also be available. Depending on the circumstances of the termination of employment, different levels of urgency should be applied in removal of access permissions and return of assets – there should be an agreed approach for urgent or exceptional cases where a leaver is considered high risk.

Another aspect for the auditor to consider is the handling of responsibilities or duties that continue for a period after employment has terminated. The auditor should check whether such responsibilities have been included in the contract.

Finally, the auditor should confirm that the responsibilities, controls and procedures applied in case of employment termination are also applied when an employee's or contractor's role changes. These should also be documented and verified.

6.6 Confidentiality or non-disclosure agreements (ISO/IEC 27001, A.6.6)

"Confidentiality or non-disclosure agreements reflecting the organization's needs for the protection of information shall be identified, documented, regularly

> *reviewed and signed by personnel and other relevant interested parties."*

Implementation guidance

Suitable confidentiality agreements should be in place before giving anyone access to confidential information, signed by all applicable parties and dated. This should be implemented for all employees, as well as any relevant external personnel, and any other organisation with which information is exchanged. The organisation should develop a set of specialised confidentiality agreements that address its requirements. Examples of what might be included in confidentiality agreements are given in ISO/IEC 27002, 6.6. This is not a definitive list but can be used as a starting point.

Requirements for confidentiality agreements can be identified by looking at the following.

- Identified legal, regulatory and contractual requirements – if these requirements impose confidentiality, e.g. as data protection legislation does for personal data, a confidentiality agreement(s) might be useful.
- Information exchanged with other organisations – confidentiality agreements with these organisations should be in place to ensure that the confidentiality of the information exchanged is not compromised.
- Asset valuation – whenever the results of the asset valuation have shown that an asset has higher confidentiality requirements, all people having access to the asset should sign a confidentiality agreement.

- Unplanned access to confidential information – it might be the case that people have unplanned access to confidential information, such as cleaning personnel in an office where desks have not been cleared. As the requirements for confidentiality agreements may change, it is important that a review process is in place that identifies new requirements and ensures that these requirements are addressed in the relevant confidentiality agreements.

Auditing guidance

Auditors should confirm that confidentiality (or non-disclosure) agreement(s) are in place, and check whether they address the identified business, legal and contractual, and information security requirements. It might be helpful to consult the risk assessment results to see how these relate to the different clauses of the confidentiality agreement(s).

Auditors should also check that any confidentiality agreement uses legally enforceable terms, so that it is valid in disputes that might arise, and that the organisation has carried out due diligence to ensure that confidentiality agreements are not in conflict with any existing applicable legislation and regulations. Specialist expertise might be required to review legal documents.

The auditor needs to confirm that there are defined review and updating procedures in place for each of the confidentiality agreements, and there should be processes in place that ensure that appropriate agreements are signed before access to any confidential information is given (see also 6.2).

6.7 Remote working (ISO/IEC 27001, A.6.7)

> *"Security measures shall be implemented when personnel are working remotely to protect information accessed, processed or stored outside the organization's premises."*

Implementation guidance

As in the mobile computing environment, the main security problems with remote working from a fixed address (similarly to working from a mobile or temporary location) arise from the location. The user's home does not normally have the same level of physical security, and their work area is often easily accessible by family members and visitors. A coffee shop will not have any physical security, and there is the potential for members of the public to see information on paper or on a screen. To reduce these risks, remote working should only take place after the organisation has developed appropriate policies and procedures, put in place physical controls to secure the work area (as possible), and raised the awareness of the employees doing remote working sufficiently to control the physical and logical access to the information processing facilities used for remote working activities.

Connections between the organisation's site and remote working facilities should be secured to ensure that information cannot be destroyed, damaged, compromised or modified. Information that is accessible remotely should be restricted to that which is required for the tasks being performed.

ISO/IEC 27002, 6.7 contains a detailed list of actions the organisation should consider before authorising any remote working activities, and security controls that may be relevant.

Auditing guidance

A list of users who are authorised to carry out remote working should be recorded, along with what types of activities are authorised, in what locations, what data they are allowed access to, and what controls should be in place. It should be possible for the organisation to show how it verifies that these controls are in place, perhaps via a register, along with a risk assessment for remote working activity. It should be clear who is responsible for implementing and maintaining which of the controls (i.e. the user or the organisation). Remote working activities should only be authorised if sufficient controls are in place, including physical controls, access controls and verification of security of the connection and the remote working equipment (also see 7.9). The remote equipment should be included in the asset register. There should also be some mechanism for establishing and controlling what information is transmitted to, from and used at home, or other remote working environment.

There should be a defined policy on the use of the equipment for non-work-related activities, such as games software, accessing the Internet, etc., any of which can introduce problems when allowed to interfere with sensitive data. ISO/IEC 27002, 6.7 provides a useful list of what the auditor can review to determine whether the organisation has adequately secured remote working.

6.8 Information security event reporting (ISO/IEC 27001, A.6.8)

> *"The organization shall provide a mechanism for personnel to report observed or suspected information security events through appropriate channels in a timely manner."*

Implementation guidance

The definition of an information security event is often difficult in practice, and clear guidance and training (also see 6.3) is required to ensure that all staff can recognise one. In plain terms, an information security event is anything that could result in loss or damage to information or assets associated with information, or an action that would be in breach of the organisation's security procedures.

If information security-relevant events occur without being reported and responded to, they might cause more damage than necessary and present a lost opportunity to prevent recurrence. Failure to report events also gives a false sense of security and can bias risk assessments. Without a reporting procedure, even a major event might not find its way to those responsible for investigation and recovery until serious losses have been experienced. Minor events might also be reacted to and recovered from without a weakness in a control being recognised and corrected.

Procedures should require all users to note and report any observed or suspected security weaknesses in, or threats to, physical environments, systems or services. Build a culture of 'no blame' incident reporting. If staff are blamed for their mistakes, they will be tempted to cover up the problems.

Users should report these matters either to their line management, directly to their service provider or to any other defined point of contact as quickly as possible. ISO/IEC 27002, 6.8 provides a list of categories of information security events. It is important that all employees, contractors and third-party users understand that all types of security events should be reported, not just those related to IT – an open window, or a paper left on a train, may also be relevant. The report should then be recorded and investigated. Users should be aware that they should not try to exploit any identified weaknesses in any way.

A number of events might already be reportable under the procedures of other departments. Failures of computer and telecommunications equipment, for instance, will be reported to engineers for repair. However, they should also be reported and recorded as information security events (loss of information and service availability). Ensure that there are procedures covering the reporting and investigation of events, and that resolving actions are tracked and reported on. Reporting procedures should include standardised forms, and guidance on initial actions to take (or avoid taking). Contact points should be documented and staff made aware of them.

Timescales for reporting are also very important, as customers or legal requirements may state a minimum time for the organisation to notify them after it becomes aware of an incident. The organisation should ensure that sufficiently robust criteria are used to validate an initial report to avoid false alarms (see 5.25).

An important consideration is what to report externally, when and to whom. The organisation should define the circumstances that would lead to notification of third parties or the public, based both on the scope and impact of an incident, and on the legal and contractual frameworks that are applicable in each of its geographies and to each of its customers.

Auditing guidance

The auditor should confirm that the organisation has appropriate procedures and management channels for reporting information security events. Auditors should check that the procedures deal with all possible events and provide an identified point of contact and sufficient response. If an organisation claims to have had no events to report, and thus the process cannot be demonstrated, it is likely that events and problems took place but went unnoticed. The absence of reports does not evidence a well-functioning information security event reporting procedure.

A selection of personnel should be interviewed to verify that they are aware of their responsibility to report information security events or weaknesses; they should know who the points of contact are, and how to raise a report. The auditor should check that the reporting forms are easy to find and fill in, and that they record information that is relevant to the incident management process. The importance of timely reporting should be emphasised both in training and awareness materials and in general documentation provided to personnel. The procedures for reporting should include requirements for employees to not attempt to exploit security weaknesses, for example to gain unauthorised access – even if the intent is just to prove the weakness, this might cause damage or break the law.

The auditor should check that the definition of what is and isn't an information security event is clear, and that staff understand this. It may be useful to ask example questions such as, 'Would you consider finding an open and unattended security safe a security incident?' and 'If somebody reported receiving somebody else's salary slip, would that be a security incident?'. Obviously, such questions need to be tailored to the context, but answers from staff can be revealing and indicate the general approach to such matters.

Where reports are present, check the response to the event. Has it been resolved? Have the root causes been investigated (also see Clause 10.2 of ISO/IEC 27001)? Has the person providing the original report been informed of the outcome (if this is not confidential)? Are procedures in place to address failure to report information security events?

CHAPTER 7: PHYSICAL CONTROLS

7.1 Physical security perimeters (ISO/IEC 27001, A.7.1)

> *"Security perimeters shall be defined and used to protect areas that contain information and other associated assets."*

Implementation guidance

Premises that contain business processes, information, services, IT and other assets are vulnerable to unauthorised access and undesirable activities. Persons attempting such activities might work for the organisation, so internal protection should be considered as well as perimeter protection (think pomegranate instead of coconut); this is an example of where the concept of 'zero trust' can be productively applied.

Small premises might comprise a single physical location with just one perimeter. Larger premises might need to use several perimeters, and hence be divided into multiple zones. It is important to specify and describe the perimeter of each zone properly. It is also important to think in terms of methods of breaching perimeters – for example, if the perimeter of the organisation is an internal wall (as in the case of shared premises), does the wall go all the way up to the ceiling, or only up to the suspended ceiling?

The objective is to be able to control entry into (and possibly exit from) every zone, and additionally to record entry to, and exit from, sensitive areas. A security model can be prepared showing, perhaps schematically, the various zones and the connections between them. A risk assessment should be used to identify appropriate perimeters, and to select controls to give adequate protection. Procedures should be defined to control the management of physical security. Give due consideration to out-of-hours working, lone working and any necessary authorisation, supervision and monitoring. The implementation guidance in ISO/IEC 27002, 7.1 contains suggestions for physical security perimeters.

Auditing guidance

The organisation should be able to explain what perimeters are in place, and what they are intended to achieve (e.g. separation of zones handling data of different classification, or of zones occupied by different customers). Auditors should check how access to the protected environment is controlled in normal and emergency circumstances (e.g. when evacuating following a fire alarm) and whether the security measures that have been identified are sufficient for the needs of the organisation.

To assess the physical protection in place, auditors will need to look for opportunities for access, and verify that they are managed correctly. Locked fire escapes, closed fire doors and attended reception areas are all examples of correctly implemented perimeter security controls. Conversely, incomplete perimeters can undermine a security plan (e.g. where there is a master key held by a building management company, and no processes or risk assessment to ensure that this approach maintains the perimeter of the organisation).

The implementation guidance in ISO/IEC 27002, 7.1describes several different issues that should be considered and implemented for the security perimeters of an organisation.

7.2 Physical entry (ISO/IEC 27001, A.7.2)

> *"Secure areas shall be protected by appropriate entry controls and access points."*

Implementation guidance

A secure area, in this context, is any area that the organisation identifies as requiring access control to limit the roles that can enter. Such areas can include the entire premises, but certainly server rooms, network equipment rooms and plant rooms (power, air conditioning). A clerical area handling sensitive data (such as sales calls, customer service or banking) might also fall into this category, as may delivery and loading areas. Different secure areas may need different levels of security and access control. Secure areas, and the protection to be provided by the controls within such areas, should be determined by a risk assessment. Decisions on these matters should be treated as equivalent to protecting access to electronic systems via login; the physical access route should not be the 'weakest link'.

Potential threats may include breaches of confidentiality, unauthorised tampering with equipment (causing loss of integrity) and equipment theft (resulting in loss of availability).

Appropriate entry controls might include a check of staff ID cards, the use of biometric verification (e.g. via fingerprint) and/or the entry of a password or PIN. All those accessing secure areas should be appropriately checked, and badges should be consistently used to identify authorised personnel. The organisation should also ensure that visitors are registered and escorted, and that any person not wearing an identification badge is reported to security personnel. If staff habitually do not wear passes, then a visitor may 'become' a member of staff simply by removing their visitor badge. Further specific controls are listed in the implementation guidance in ISO/IEC 27002, 7.2.

Breaches of confidentiality, integrity and availability can also occur through delivery and despatch. There are specialised threats relating to malicious delivery (e.g. letter bomb) and unauthorised despatch (e.g. theft by replacing delivery addresses on a shipment of files).

A busy organisation will experience many deliveries and collections. No one will be surprised to see packages being delivered or collected by strangers (delivery staff). It is therefore essential to control this activity to ensure that deliveries are expected, that collections are properly authorised and that third-party delivery staff are properly controlled with respect to access. In addition, the delivery and loading areas might be easily accessible by the public, and as there are people coming and going, someone might use the chance to sneak into the organisation's premises.

To control these problems, physical zoning with access controls is recommended to isolate delivery and loading from the most secure areas and restrict access from outside to identified and authorised personnel. Internal procedures should be used to ensure that the transfer of goods between the loading bay and secure area is controlled, and that the incoming goods are inspected for potential threats. Complete records of all deliveries and despatches should be kept, and the ingoing and outgoing material should be reflected in updates of the asset inventory (see also 5.9). The names of all delivery drivers and vehicle numbers should be recorded.

Auditing guidance

Auditors should check the entry controls in place, ensuring that they restrict physical access to authorised people only. Do employees wear badges and is this mandatory? What about visitors – are badges issued, is their entry and exit logged, and what restrictions are placed on their movements? Are persons without badges routinely challenged? Auditors, invariably being visitors to the organisation, can determine this first-hand (e.g. by quietly removing their badge and awaiting developments).

Auditors should also check the audit trails of access and ensure that procedures for the review and update of physical access rights are in place. Authorisation, in terms of access rights and restrictions, might be in a variety of forms. It could be described in job descriptions, it could be written into procedures or it could be listed at the point where the restrictions apply, such as on a label affixed to a door. Auditors should take a view on the appropriateness of each approach in the specific organisational context (e.g. could a sign on a door be easily amended or replaced?).

The auditor should check that the risks relevant to loading and delivery areas have been identified by the risk assessment and security procedures, and that adequate measures have been taken to both prevent and mitigate potential security breaches. For example, who receives goods: the person requiring the goods, a stores employee, or a general receptionist? What happens to the goods after receipt: are they sent directly into the secure area? Are they held in a store? Are they left on someone's desk? How are relevant items added to the asset inventory (see also 5.9) upon receipt, and is the asset inventory updated when information or associated assets leave the organisation? How is access to the delivery and loading area controlled? Is it possible for the public to gain access, and what controls are in place between the delivery and loading area and other parts of the organisation?

7.3 Securing offices, rooms and facilities (ISO/IEC 27001, A.7.3)

> *"Physical security for offices, rooms and facilities shall be designed and implemented."*

Implementation guidance

The organisation should identify the controls required to secure offices, rooms and facilities. These controls should be appropriate to the value, liability and importance of the information and related assets within the areas. A risk assessment should support the decisions made.

It may be appropriate to avoid giving clues about the location of potential targets, for example by using signs pointing to the server room or indicating the purpose of rooms or buildings. For the same reason, directories and internal phone books should not be accessible by anyone outside the organisation.

Electromagnetic shielding should be considered only if the organisational context merits it – the consequent lack of ability to reach mobile phone networks may impede normal business operations, and may also affect business continuity planning.

The risks of loss of confidentiality, integrity and availability all increase as more of the organisation's key information is centralised. These concentrated repositories become critical to the organisation. Effective security is then especially vital, both outside and inside, to ensure that losses are not experienced.

Auditing guidance

The level of protection provided for a secure area should be consistent with the most sensitive and critical information held in that area, and consistent with the organisation's documented procedures for the handling of classified information. There is a clear link here to risk assessment. Auditors should confirm that the information security requirements have been identified, and that the protection in place is adequate for this.

A list of security controls that might be applicable to protect secure areas is given in the implementation guidance in ISO/IEC 27002, 7.3. Auditors should also try to identify the use and purpose of rooms where information is processed, or try to get access to internal telephone directories/lists, to test these controls. Phone lists are commonly left next to phones at reception, or on notice boards in kitchens. Entry codes for doors may also be written on notice boards or whiteboards.

7.4 Physical security monitoring (ISO/IEC 27001, A.7.4)

> *"Premises shall be continuously monitored for unauthorized physical access."*

Implementation guidance

Just as networks are monitored for evidence of intrusion, so too should physical premises be monitored for break-ins. This is a well-known control used by many private individuals and organisations as a matter of course (e.g. home burglar alarms linked to a monitoring service), and can be implemented via manual methods, automated tools, or a hybrid approach. Procedures should be defined to cover what is being monitored, how, and why. Processes should also clearly state what happens, and who in the organisation will be notified, if an alarm is tripped (see ISO/IEC 27002, 5.24). In addition, define what should be done to preserve appropriate levels of information security in the event of a physical security breach (see. 5.29). Finally, a good and resilient approach should be taken to minimise false alarms, as these can lead to slow and inadequate response when a real incident arises – and some monitoring services will either terminate their service or start charging inordinate fees for repeated false alarms.

Auditing guidance

Monitoring systems, personnel or services should be well documented, and their effectiveness checked by intermittent tests of the systems in question (much as fire alarms are tested). Interview staff responsible for monitoring to determine whether they know what to do if an alarm goes off – how do they verify a false positive? How quickly do they need to escalate? Is there a clear chain of notifications, and do the persons to be notified know how to respond? Compare this with the monitoring procedures and processes for incident management (see ISO/IEC 27002, 5.26). When do roles responsible for information security get involved if an alarm goes off? If they are not included in response activities, then information security incidents relating to physical information theft or damage may not be managed correctly (see 5.29).

7.5 Protecting against physical and environmental threats (ISO/IEC 27001, A.7.5)

> *"Protection against physical and environmental threats, such as natural disasters and other intentional or unintentional physical threats to infrastructure shall be designed and implemented."*

Implementation guidance

Organisations are always vulnerable to threats outside their control. The selection and design of the site and the controls applied should take account of the possibility of damage from fire, flooding, explosion, chemical leak, civil unrest, and other forms of natural or man-made disasters. Consideration should also be given to any threats presented by neighbouring accommodation. For example, are hazardous materials being handled? Do any neighbours carry out business that may attract hostile attention?

The selection of controls should be carried out in consultation with specialists in the relevant areas, documented as required in Clause 6.1.3 of ISO/IEC 27001, and the necessary training recorded in staff training records. The fallback arrangements and backups taken should conform to the business continuity plan (see also 5.29).

Auditing guidance

Auditors should check the provisions the organisation has in place to react to natural and man-made disasters, and the physical protections in place to limit the damage. Records should exist of specialist advice pertaining to the main hazards that are likely in the circumstances, and should identify measures that are to be taken in the event of a hazard being realised. Secondary effects should have been considered, for example the ability of a server room to retain adequate climate control in the event of a power failure.

Have any reasonably predictable sources of hazard been omitted? For example, if a neighbouring site poses a threat to the organisation, has this been considered? The auditor should also investigate whether these arrangements link in with, and conform to, the business continuity arrangements the organisation has implemented (see also 5.30). Another issue to consider is the emergency support and environmental protection in place. Has the organisation assessed whether there is a fire hazard, or whether the site could be flooded – and what is there to prevent or mitigate these dangers?

Auditors should check if measures have been implemented as described. It might be helpful to walk through the site to identify weaknesses, such as large quantities of paper stored

in an aisle without specific protection, inaccessible fire
extinguishers, a server room in the basement, lavatory waste
pipes directly above a critical server rack, etc. Some
measures may be eroded by time (e.g. fire extinguishers
being used to prop fire doors open on hot days).

7.6 Working in secure areas (ISO/IEC 27001, A.7.6)

*"Security measures for working in secure areas shall be
designed and implemented."*

Implementation guidance

In addition to setting up physical perimeters, applying entry
controls and securing offices, rooms and facilities for day-to-
day operations, the specific security requirements of areas
involving sensitive work need to be considered. For
example, an organisation could be working on a new
product, the design of which has high commercial value and
is ahead of its competitors. Another example might involve
a project or process that needs to be protected from damage
or unauthorised modification.

The work in secure areas should be protected and supervised
as described in the implementation guidance in ISO/IEC
27002, 7.6. Which measures are applied, and the degree of
protection they will afford when used in concert, should be
determined by a risk assessment based on the work going on
in the secure areas, and the protection requirements of the
information and other assets in these areas.

Auditing guidance

Personnel working in secure areas should be subject to specific measures that ensure sufficient security is implemented for the sensitivity and criticality of the information that is processed in such areas. Auditors should check:

- that entry controls are in place to ensure that only authorised personnel have access to secure areas;
- to what extent the work going on in such areas is generally known, and whether this exceeds any rules on 'need to know';
- how easy or difficult it is to take information (e.g. in the form of paper or electronic media) in or out of such areas;
- whether it is possible to take mobile phones, or other photographic, video, audio or recording equipment, inside such areas, and to use it, or leave such equipment there to record;
- whether the work in such areas is sufficiently supervised; and
- whether mechanisms are in place to ensure that dual controls (where the presence and simultaneous activity of two individuals are required to authorise an action) are applied where appropriate.

The auditor should also check that the procedures for working in secure areas are applied consistently to everyone in those areas, including employees, contractors and third parties.

7.7 Clear desk and clear screen (ISO/IEC 27001, A.7.7)

> *"Clear desk rules for papers and removable storage media and clear screen rules for information processing facilities shall be defined and appropriately enforced."*

Implementation guidance

Offices, especially when open plan, provide good opportunities for people to walk around and read documents or information on screens that they are not authorised to view. Such people might include other staff or other individuals, such as visitors, air-conditioning engineers and cleaners. Most people have a camera in their pocket (as part of their mobile phone), so photographing documents on paper or on a screen is very easy and unobtrusive. If people leave material visible on screens when they leave their work area, or leave materials on the desk, this might lead to unauthorised persons obtaining sensitive information.

A disorderly desk may lead to loss of documents due to misfiling, or even putting them in the wastepaper bin by mistake, which could cause sensitive information to be viewed by unauthorised persons (such as recycling contractors). Information left out on desks is also more likely to be damaged or destroyed in a disaster such as a fire, a flood or an explosion. In addition, paper, being flammable, may contribute to a fire hazard.

Printers and fax machines may also pose a risk, and the use of 'pull printing' (where the user has to authenticate to the printer before their job is printed) may be appropriate, as long as users know not to wander off while a particularly long document is being printed. Fax machines expected to receive sensitive data should be in a secure location, and regularly checked for messages.

Organisations should adopt a clear desk policy for papers and computer media, and a clear screen policy for information processing facilities, to reduce these risks. Staff usually see this as an onerous control, so training should emphasise the benefits of working in an organised and tidy environment, and mandate that screensavers with passwords are used, or equipment is switched off when leaving the office. Compliance should be monitored and rewarded, persistent offenders noted, and training and/or sanctions applied.

Auditing guidance

The objective of this control is both to ensure that sensitive information in any form (processed electronically, on paper or media, etc.) is not left unattended, and that information is not lost or made available to unauthorised people. This should apply to both working and non-working hours. Controls applied should be matched to the classification of information (see also 5.12).

The auditor should confirm that the organisation has a policy in place to prevent sensitive information being accessed by external individuals, e.g. cleaning staff. The auditor should also check what happens when desks, filing cabinets and safes are left unattended during the day, and when cleaning staff or other visitors (such as the auditor) enter the office. Furthermore, is there a process for warning staff who may be working on sensitive data, so that they can clear their screens and/or desks before the visitor enters? The auditor should check the risk of access to computers while staff are absent (irrespective of the duration of this absence): password-protected screensavers, switching the computer off or any other form of clear screen control should be applied.

Are faxes and printers suitably located, is there a pile of printouts on the printer and/or fax machine? Is the fax machine in a secure location if it is intended to receive sensitive material? Is pull printing in place or available for use for sensitive documents? Do staff know how to invoke it if it is optional?

Where necessary, additional access controls should also be in place. If the whole area is covered by an appropriate level of security, and all staff are appropriately authorised, then additional measures might not be needed. Check that the overall policy is clear, and that staff are aware of and follow the appropriate procedures. Sanctions for non-compliance should be simple, fair and consistently applied.

7.8 Equipment siting and protection (ISO/IEC 27001, A.7.8)

"Equipment shall be sited securely and protected."

Implementation guidance

Equipment can be vulnerable to damage and interference, with a resultant loss of integrity and availability. Inappropriate accessibility can lead to unauthorised use and/or breach of confidentiality of the information available.

Physical damage can arise from poor environmental conditions, particularly in industrial situations where moisture, vibration, heat, dust and chemicals can all take their toll. Electrical and electromagnetic interference can also be significant in some environments, and should be tested to identify possible problems resulting from interference. This can be at an industrial or very localised scale; for example, a person may be ornamenting their processor case with fridge magnets. Staff training in proper use of equipment is thus basic to its protection.

It is relatively easy to protect equipment such as communications devices and connection panels from inappropriate access – simply lock them in a small room or equipment cupboard with appropriate temperature and humidity controls. Equipment required by operating staff should be available in their workspace and special protection, such as keyboard covers, can help protect it. Ensure that the risk assessment covers this kind of situation and identifies solutions for equipment requiring special protection.

Other problems relate to the people working with the equipment; for example, equipment can be damaged if eating, drinking or smoking takes place too close to it. There should be a policy in place to prevent this. In addition, there are risks associated with equipment displaying confidential information. There are different ways to handle this, including a clear desk and clear screen policy (see also 7.7), rules for unattended equipment (see also 8.1) and restriction of viewing angle to avoid information being viewed by unauthorised people walking past.

Where networked equipment is used, remember that remotely accessible equipment probably requires more security attention than in-house equipment. Clearly establish the extent of the organisation's responsibilities for the network, and apply appropriate protection at the boundaries. Ensure that remote equipment is accounted for in inventories, security scope and risk assessments.

A list of guidelines is provided in ISO/IEC 27002, 7.8.

Auditing guidance

The organisation should be asked to demonstrate how its equipment is protected from environmental threats and hazards, and opportunities for unauthorised access. Equipment should be sited away from risk areas, such as windows that could be easily broken during a burglary without setting off an alarm. Consider also that terminal screens might be viewed from outside a protected area, and information can be leaked through electromagnetic or other emanation (e.g. the sound of keystrokes); these topics should be covered in risk assessments if the risk is plausible and relevant.

In some environments, it may be appropriate to secure computer equipment to desks. As well as malicious damage, equipment needs to be protected from accidental damage from a very untidy or poorly managed environment, unrestricted access, unstable racks, spilt drinks, etc., and from environmental hazards such as water, chemicals and fire, and electromagnetic interference (see also 7.5). Check that such measures have been considered, a risk assessment has taken place, and that adequate protection is implemented.

Looking beyond the immediate computer area, does a fire or water hazard exist in adjacent areas? A large organisation will probably have a site layout plan; this may be used to identify risks that may spill over into the organisation's scope.

7.9 Security of assets off-premises (ISO/IEC 27001, A.7.9)

> *"Off-site assets shall be protected."*

Implementation guidance

In many organisations, staff can regularly be expected to take equipment, data and documents away from the premises. This might be to work at home, or to attend meetings at other premises. The use of assets outside the organisation's secure environment increases the flexibility of the organisation, but adds complexity to the protection of its information and associated assets. The impact on the organisation's risk profile will largely be determined by its activities and risks. This is further complicated by the fact that, if bring your own device (BYOD) policies are in place, the device may not be under the control of the organisation, but the information held on it or accessed from it may belong to the organisation. Portable/mobile devices are particularly vulnerable to theft when in public places, leading to breaches of confidentiality as well as the loss of the device.

There are three main alternatives that an organisation may consider to address the risk of theft or misuse, as follows.

1. Removal of all assets containing sensitive information is prohibited. On the face of it, this is the simplest approach, but difficult to implement for most organisations. Highly restricted environments might need to use this approach.

2. Removal of assets containing sensitive information is permitted with appropriate controls. The organisation needs to be very clear what information is involved and what controls are needed.
3. Removal of assets containing sensitive information is permitted without controls. This can be very dangerous, and should always be accompanied with additional controls regulating the handling of sensitive information when it is outside the organisation's premises.

Equipment, information and software, etc. should not be taken (or transmitted) off-site without formal authorisation. It is essential that the organisation knows where its assets are and who has responsibility for them. All items of equipment should, where possible, be marked to indicate their ownership.

Those carrying items such as laptops, other mobile devices and sensitive business information (in digital form or on paper) in and out on a regular basis should be provided with a method to demonstrate their permission to carry these assets with them, which should be available on demand. Additional verification and authorisation should be considered where an individual needs to carry information, or devices capable of accessing information, to another jurisdiction that has different laws on information access and protection.

Where items are on long-term loan, for instance to home workers, the individual should be required to attest periodically that the items are still in their possession, in good condition and necessary for their work. Procedures should be implemented to ensure that those leaving employment return all company assets before departure. Alternatively, where the organisation allows staff to retain or buy their equipment, procedures should be followed to permanently erase any stored organisational information before the individual is given the device (see 7.10).

The risk assessment should aim to ensure that the security provided off-site is equivalent to the security arrangements on-site, and appropriate insurance arrangements should be implemented for information and organisation-owned equipment when off-site. Should it be impossible, or impractical, to achieve an equivalent level of risk mitigation off-site, the correct roles in the organisation (i.e. top management) are responsible for making an informed decision on whether to tolerate this heightened level of risk, given the potential rewards of mobile working.

The security of mobile equipment is also discussed in 8.1.

Auditing guidance

The auditor should look for evidence that the risks posed by off-premises information and associated assets have been assessed, where this is applicable to the organisation. The auditor should then check that the controls implemented for the physical protection of equipment outside the premises provide security comparable with what is implemented on-site, or that a suitable role has signed off on the difference in risk (this acceptance should be reviewed at appropriate intervals). Procedures and guidelines should be in place to ensure that equipment off-premises is not left unattended and unprotected. Where relevant, sufficient insurance should have been taken out. Verify the excess payable and the requirements on the organisation to ensure that the insurance will pay out when required.

Additional protection mechanisms are also described in 8.1, which addresses mobile computing, and 6.7, which covers the security of remote workers.

7.10 Storage media (ISO/IEC 27001, A.7.10)

> *"Storage media shall be managed through their life cycle of acquisition, use, transportation and disposal in accordance with the organization's classification scheme and handling requirements."*

Implementation guidance

Media containing organisational data presents inherent risks related to the loss of confidentiality and availability. For example, newly acquired USB sticks from online retailers may contain pre-installed malware, posing immediate security threats. Effective management of various media types – including, but not limited to, backup tapes, disks, USB sticks, removable hard drives, CDs, DVDs and printed media – is imperative.

The organisation should develop and implement comprehensive procedures to ensure that media is procured, used, maintained, transported and disposed of in a manner that aligns with the organisation's information classification scheme. Adherence to supplier recommendations regarding storage conditions is advised, along with considerations for media deterioration and obsolescence.

The unauthorised removal of assets may indicate potential theft, leading to risks associated with non-availability and loss of confidentiality, especially when the media contains sensitive information or readable software. Risk assessments should account for the ease of transportability of small media items in and out of premises.

The transportation of media entails risks of loss, unauthorised access and misuse, thereby affecting the confidentiality, integrity and availability of the information or software contained within. Using risk assessments to determine appropriate transport methods and controls is crucial. This may include secure parcel delivery, personal delivery by verified couriers, and the selection of appropriate courier services and packaging solutions, such as locked containers or tamper-evident packaging for sensitive or high-value items. Strategies such as dividing shipments and employing data encryption are recommended to significantly mitigate confidentiality breaches in the event of loss or theft. However, it is critical to ensure that encryption keys are not transported with encrypted data (also see 8.24).

All dispatches of media should be recorded and authorised. Media perceived as obsolete or redundant may still contain valuable information, and its improper disposal can lead to significant confidentiality breaches. This is particularly pertinent when disposing of hard drives, disks, tapes and paper files. The return of damaged media to manufacturers for repair or disposal also presents data recovery risks by unauthorised parties.

Procedures for handling media containing classified information should include specific methods for secure destruction and disposal. The organisation should establish formal procedures for the disposal of media that is no longer required. For detailed procedural guidance, refer to ISO/IEC 27002, 5.10, applying these controls as necessary based on the significance and sensitivity of the information stored on the media. A suitably detailed record of the destruction of sensitive items should be maintained.

Those bringing property into secured areas should be required to log the property on entry so that they can leave with it without difficulty. Appropriate documentation should be kept regarding procedures, authorisations, off-site inventory and returns.

Auditing guidance

The auditor should review the organisation's policies describing which media should be used. Are staff permitted to use any media they like, or is there a requirement for approved media to be used; and how has this decision been arrived at (look for a risk assessment)? For media handling, check records to verify that these procedures are followed and interview random staff to establish that everybody is aware of them. The handling and storage should be appropriate to the classification of the information stored on the media, and the media should also be stored and disposed of in accordance with manufacturer recommendations. Media should also be replaced before it is obsolete, and included in an asset inventory (see 5.9).

Another aspect the auditor should check is how media is removed from the site. This might be for transfer to secure archive storage, by personnel for business use, or for destruction. There should be a well-defined procedure and logging mechanism in place, as well as authorisation required, in each case as appropriate. The procedure should ensure that removable media, if no longer needed, are erased to ensure that no information is leaked. If media contain sensitive information, the auditor should check how they are labelled and handled, and confirm that there are procedures for them to be destroyed or erased before being reused or discarded.

The auditor should also confirm that the transport arrangements for various media afford sufficient protection. Whatever controls are in place, this is a difficult area to police. Check that the organisation has properly identified this in its risk assessments, and what compensating controls have been applied.

The auditor should confirm that whenever information is physically transported, the organisation has considered what protection is in place to protect the media holding the information. What are the transport arrangements? If couriers, do they have secure and tamper-proof containers? Have they been identified as trustworthy couriers? Data might be transmitted by staff on disks or tapes or perhaps on mobile devices – are these devices secure enough for the information carried? Who determines the method of transportation? What criteria do they use? Where couriers are employed, the methods of transportation might be the carrier's default methods: are these sufficient? Consider also postal services: are these secure?

In all cases where there is a requirement for secure transport of information, the auditor should ask to see the procedures defining these arrangements, and a sample of the records. Records should include details of what was delivered, when the courier picked it up, who authorised the transport, and when the delivery reached its destination.

The auditor should check which approach the organisation takes to the removal of sensitive information, and then look at the documented procedures for managing the risk of this removal (if it is permitted at all). Is a booking-in/out system in use? What authorisation is needed and recorded? Is this for all items or only a restricted range? How does management monitor compliance? Are spot checks carried out? Does the confidentiality agreement (see 6.6) cover responsibility for information held while off-premises?

The auditor should confirm that formal procedures are in place for asset disposal/destruction. What arrangements are there? How do external contractors handle them? Check that the organisation has carried out proper security and process checks, and that the most sensitive level of information handled in this way is known and verified. The auditor should check that – whatever the specific arrangements are – sensitive information cannot be compromised through the disposal process, because it has already been erased or destroyed. There should be a logging process for the disposed-of media, and what has been done to it before disposal. Check that this provides a satisfactory audit trail.

There should also be a process to govern the handling of damaged media containing classified data, which includes a risk assessment and determines whether the media will be destroyed rather than sent for repair.

7.11 Supporting utilities (ISO/IEC 27001, A.7.11)

> *"Information processing facilities shall be protected from power failures and other disruptions caused by failures in supporting utilities."*

Implementation guidance

Supporting utilities, such as electricity, heating/ventilation and air conditioning are a prerequisite for the use of any computing and communications equipment. In addition, the loss of facilities like clean running water, while it does not directly affect equipment, is likely to result in site closure (on health and safety grounds). While many organisations take a reliable public supply of electricity or water for granted, they are still at risk of disruption resulting from incidents, such as the activities of someone with a digger – no supporting utilities usually means no availability.

The need for supporting utilities should be identified. The utilities should be regularly inspected and tested to ensure their reliable functioning. The water supply should be sufficient to guarantee air conditioning as necessary, as well as sufficient fire protection. A problem with any of the essential supporting utilities should be identified by an alarm system.

The risk assessment should identify facilities that require redundant power supplies, especially those supporting critical business operations – and their environmental management systems (a server room without functioning air conditioning will overheat and shut down surprisingly quickly). The selected option, such as an uninterruptible power supply (UPS) or generator, should be capable of sustaining sufficient power for the maximum potential period of a power cut, or at least for the time identified in the business continuity plan. Some equipment requires a very stable electrical power supply, free of peaks and troughs (spikes) in power. If this requirement is not met, power spikes can lead to a loss of availability through equipment damage or failure.

> *NOTE: Building management systems in the scope of the organisation should be treated as for other IT systems with regard to malware protection, backups, vulnerability management, audit, etc.*

Auditing guidance

The level of protection that should be provided to safeguard an organisation from disturbances in its supporting utilities depends on the security requirements of the information held in each environment. For example, information with high availability requirements should be protected by controls designed to ensure sufficient supporting utilities. Auditors should check that the organisation has considered all necessary supporting utilities, and has implemented controls to ensure adequate levels of service, both technically and via agreements with external service providers (e.g. telecommunications providers and utilities). Auditors should also check that there are procedures in place to inspect and test all supporting utilities regularly, for example by asking to review the records of those tests.

For higher power requirements, check that sufficient facilities such as standby generators, UPS units, redundant disk (RAID) units, etc. are in place. If this is the case, look more closely at the power supply support. Does it have sufficient capacity to cover air-conditioning requirements? What is the extended operating period? Does it match the documented requirements? How is this verified? Is equipment maintained and tested in accordance with the manufacturer's recommendations? What actions are taken to detect malfunctions? The auditor should also check that emergency lighting is provided in case of a power failure.

The auditor should also check that there are redundant connections to key utility providers if appropriate (or that this measure has been considered and agreed to not be necessary), to prevent failure of one connection resulting in the loss of a critical service.

7.12 Cabling security (ISO/IEC 27001, A.7.12)

> *"Cables carrying power, data or supporting information services shall be protected from interception, interference or damage."*

Implementation guidance

Unless such cables are properly installed, it can be very easy to damage them, especially their connectors, leading to a loss of availability. It can sometimes be difficult to trace the fault. Cables left on floors and hanging loose around walls are a health and safety hazard and will suffer excessive wear or pulling, leading to damage; rodents also sometimes have a nearly supernatural attraction to conduit sheathing. In addition, unclearly marked cables might be subject to inaccurate connection. Finally, including power and data cables in the same conduit may result in interference affecting the data cables.

In some organisations, communications cables might be at risk of tampering, resulting in interception of the information they carry. In this case, they should be protected by conduits, with all connections made in locked equipment rooms or boxes. While physical protection will be the principal safeguard to consider, there are also data transmission controls, such as encryption, that can be employed. The risk assessment should determine where this is necessary.

Public access to roadside telecommunications junction boxes might also pose a risk in some places, both from physical damage and tampering. Discuss this with your network service provider with a view, perhaps, to relocating the box underground beneath a secure lid. ISO/IEC 27002, 7.12 provides further guidance on cabling security, especially on how to protect sensitive or critical systems.

Auditing guidance

The organisation should be able to demonstrate that connectors and cables are adequately protected from interception, interference or damage. Are they correctly fitted and properly routed, or are they badly implemented and placed where they could be damaged or cause an accident? ISO/IEC 27002, 7.12 provides a list of guidelines that should be considered for power and telecommunication cables.

A good indication of the status of the cables is the documentation describing the power and communication lines, and the cable colour coding or labels used. Auditors should check that the organisation has considered the communication risks and looked for potential weak points regarding network cabling routed between departments or buildings, unprotected or unsegregated telecommunication and power lines or cabling accessible to interruption or eavesdropping.

7.13 Equipment maintenance (ISO/IEC 27001, A.7.13)

> *"Equipment shall be maintained correctly to ensure availability, integrity and confidentiality of information."*

Implementation guidance

The correct operation of computing and communication equipment can lead to a false sense of security. The sudden failure of equipment that has worked faultlessly for years can have a profound effect on the integrity and availability of business processes and services – especially if the equipment cannot readily be replaced. Equally, if a system does not 'fail safe' – e.g. a VPN that fails open – then confidentiality can also be at risk.

Most equipment is supplied with maintenance instructions and these need to be built into operating procedures. Ensure that maintainers are authorised and qualified, and that they are accompanied when carrying out their maintenance work. Keep records of faults and maintenance – monitoring these will help in the judgement of when equipment should be replaced and so avoid the sudden failure. Also ensure (either by deleting confidential information before maintenance activities or protecting it in other ways) that no confidential information is disclosed.

Auditing guidance

The auditor should confirm that the organisation has controls in place to ensure equipment maintenance in accordance with suppliers' recommended service intervals and specifications. In addition, simple maintenance processes such as regular cleaning of air filters, tape drive mechanisms and printers can save considerable disruption. Even mundane activities such as regular disk defragmenting on computers with spinning hard disk storage can extend the usable lifespan of equipment.

Look to see what maintenance activities are identified in the procedures, determine whether they are sufficient and check the records to ensure that maintenance activities in the past have taken place as specified in the procedures. There needs to be a formal fault reporting mechanism. Check for this, and for logs of defects and their rectification. The auditor should verify that only authorised personnel can carry out maintenance activities, that external personnel carrying out maintenance are always accompanied and that no confidential information is accessed. As far as is possible, equipment should be checked after repairs and maintenance for evidence of tampering and unauthorised modification (e.g. the installation of a wireless access point to permit future remote access).

7.14 Secure disposal or re-use of equipment (ISO/IEC 27001, A.7.14)

"Items of equipment containing storage media shall be verified to ensure that any sensitive data and licensed

> *software has been removed or securely overwritten prior to disposal or re-use."*

Implementation guidance

Serious breaches of confidentiality can occur when discarded storage media, such as disk drives, are accessed by unauthorised persons, for example after being sold on the second-hand market or left in a skip. Although files may have been deleted from media, they may remain accessible to anyone with the right tools. Copies can also be made of the organisation's software if it is not permanently deleted, laying the organisation open to charges of unauthorised distribution of copyright material. The organisation should use controls to ensure that any equipment to be disposed of or reused no longer contains sensitive information.

Some unexpected devices may also contain storage media; for example, printers may hold copies of files that they have printed over the last several months. CCTV cameras may also hold local buffered data.

It should be noted that certain storage devices (such as magnetic hard drives) may be adequately wiped using suitable programs, but that other types of devices (such as solid-state disks) may retain data indefinitely. Many types of storage device are relatively cheap. The organisation should consider complete destruction as a method of disposal for unwanted storage devices containing sensitive data. Magnetic storage is relatively cheap – much cheaper than the impact of theft or compromise of sensitive data.

Depending on the risks involved, the organisation may apply physical destruction of media, and this can also extend to hard disks.

Encrypted data may not be retrievable from discarded media in any reasonable time if the keys are suitably chosen and not stored on or with the media, but, as computing power increases, the risk of successful brute-force decryption also rises (also see 8.24). The lifetime of sensitivity of the data can be considered to determine the level of encryption required, or to decide whether encryption is an adequate control.

Auditing guidance

The auditor should check that the organisation has an effective process in place to ensure that any sensitive information is removed from equipment that is disposed of or otherwise taken outside of its control. The auditor should also check user awareness of the potential dangers. The organisation should understand that erasing files from media is not necessarily secure. The information is often still accessible. Check that overwriting tools are used where appropriate to reduce the chance of data recovery.

Where encryption is used to mitigate the risk of data leakage, the auditor should check that the level of encryption is adequately matched to the sensitivity of the information to be protected, taking into account the likely future increase in computing power available to an attacker over the usable lifespan of the information.

For magnetic media holding very sensitive information, specialist equipment (e.g. degaussing tools) should be employed to minimise residual data. The policy should extend to all media if labels on items holding sensitive data could be removed or modified before disposal, making positive identification difficult.

Consider also items sent for repair. Are there any checks to ensure that sensitive information cannot be accessed or interfered with, or is there a policy that all items sent for repair should have storage media removed or destroyed before dispatch? See also 7.10.

CHAPTER 8: TECHNOLOGICAL CONTROLS

8.1 User end point devices (ISO/IEC 27001, A.8.1)

> *"Information stored on, processed by or accessible via user end point devices shall be protected."*

Implementation guidance

The organisation should develop a user policy describing the controls that should be in place, and employees should only be allowed to use devices after they receive the policy and have had sufficient training and awareness education (also see 6.7 and 5.10). Devices should be required to meet an appropriate standard of security (e.g. be promptly updated to remove security vulnerabilities; see 8.8), and users should have clear guidance on what to do and who to inform should a device be lost or stolen, especially if it is a personal device (see 5.24).

Other security risks when using computing facilities are related to information exchange. Backups should be made regularly, and effective and frequently updated malware protection should be used (as available – some devices do not require this protection; also see 8.7) if any information transfer takes place. In the case of remote connections to the organisation's site, authentication not only for the machine but also for the authorised user should be in place to avoid such connections being exploited, e.g. by somebody who has stolen a laptop or mobile phone (also see 6.7). In addition, the user should be aware and capable of determining a suitably secure method to connect to the organisation's network, e.g. via VPN, to protect against interception of sensitive data. This will inevitably include a basic grasp of what wireless networks are to be trusted and to what extent.

ISO/IEC 27002, 8.1 describes further measures that can be applied to protect information and facilities accessed via user devices.

An unattended system left logged on to a service is vulnerable to misuse, providing concerns for confidentiality, integrity and availability. Other equipment that is accessible by unauthorised people is also vulnerable to disclosure, misuse, tampering and theft, leading to loss of confidentiality, integrity and/or availability.

Sensitive equipment, such as communications panels and controllers, should be locked away in equipment rooms or purpose-built cupboards. Desktop equipment such as computers should be locked or shut down when unattended. Risk assessment should determine the maximum time a session can be left open before it is automatically disabled or terminated. Keyboard and mouse use should be protected by a password while the computer is unattended. Screensavers with passwords should be used to hide the screen contents while unattended. Ensure that the strength of the password system is matched to the confidentiality and integrity requirements of the resources being protected.

Auditing guidance

The auditor should confirm that the organisation has identified the use of all user IT and communication devices. This includes the use of personal and work-owned mobile phones, laptops or other mobile devices inside or outside the organisation's premises (e.g. at home, customer sites, hotels, during travel or at conference venues), and any remote connections to the organisation's internal information processing facilities using such devices. As mobile computing and communication activities normally take place outside the organisation, their use will normally be difficult to audit directly.

The auditor should review the controls, rules and procedures that the organisation has in place to protect devices, and to connect to networks. User training and awareness, authorisation processes and security arrangements for using devices can be reviewed against the list in ISO/IEC 27002, 8.1.

As is feasible, audit evidence should be collected to check that all these controls are implemented correctly. Audit checks for controls should also include what the policy says about password and malware protection on mobile and fixed devices (also see 8.5 and 8.7), what is permitted to be stored locally, the use of cloud services for backups, and what additional software must be present to separate personal and work-related data. Check that there are sufficient controls in place to secure remote access, and that strong cryptographic controls are applied where necessary (also see 8.24).

Timeouts from sessions (via the connection to the resource being terminated, and/or the local computer screensaver) should be in place (see 7.7). Look at where this has been implemented, and determine whether the timeout periods are sufficient given the access level, the vulnerability and operational needs. Determine what the timeout is based on: specific use of the application, or simply movements of a mouse cursor. Check that this timeout is employed consistently at all high-risk locations (e.g. remote working environments; see 6.7). Where timeouts rely on operating system features such as a screensaver, check that the facility has not been disabled (e.g. by the user, if they have the relevant access rights).

8.2 Privileged access rights (ISO/IEC 27001, A.8.2)

> *"The allocation and use of privileged access rights shall be restricted and managed."*

Implementation guidance

Privileges are any features or facilities of information processing systems that enable the user to carry out system management activities or override access controls, such as maintaining the security system or the data management system. If privileges are uncontrolled, an increasing number of users will be using privileges, rendering pointless the carefully implemented access controls. The unnecessary allocation and use of privileges is often a major contributing factor to the vulnerability of systems that have been breached. Loss of confidentiality through exposure, loss of integrity through modification of data, and unavailability of data are typical consequences.

Privileged access to systems might be a politically challenging aspect to control. Systems engineers might try to persuade systems owners to authorise a privilege that is not really required. Privilege tends to be seen as a means to shortcut controls, and as a prize to be won and then held on to; a bit like getting to a higher level in a computer game. The fact is that most systems require very little use of privilege to manage them in a perfectly efficient manner.

Risk assessment should address not only the risk of providing privileges but also the consequences of not having them. Authorisation should be provided to staff, including those at a senior level, on the basis of sufficient business justification, which, in some cases, might need an independent expert opinion.

An important situation in which special privilege can be required is in the event of system failure. Fast recovery may require the skilled attention of a system engineer who might need to access the internals of a system and make changes that not only require elevated privilege, but also bypass controls that have been put in place to protect the system. Such occasions require their own controls, which will often be implemented after the event. It is essential that all the actions that are taken are properly logged, assessed and reviewed, and that further checks of the system are made to ensure that its security controls have been re-established.

What happens when the privileged person is not available? An emergency arrangement is required, such as a procedure to enable another systems engineer to obtain privilege out of hours. A user ID and password can be held in a safe under strict procedures for issue. This procedure should ensure that management will find out at the earliest appropriate opportunity that the emergency arrangement has been used, and should be followed up by review, as described above.

Auditing guidance

By definition, privileged access provides access to system features that are normally unavailable. The auditor should pay particular attention to system administrators, systems engineers and supplier's engineers, and others who have 'super user' access to facilities. The auditor should check that the level of access provided is appropriate to the business purpose. Full administrative access should not be used as the default approach to providing elevated privilege unless the system cannot provide any intermediate levels between standard user and full access. The auditor should also verify that more than one person has monitoring access and the ability to supervise activities.

For critical functions, such as system administration, there should be a special ID assigned to each user solely for this purpose, in addition to their standard user account on the system. Logging of usage of the account should also be in place. Interview users who have been granted elevated privileges and establish when they have used their accounts. Look at logs to ascertain that usage is being recorded. How long are these logs kept? Can they be modified? When and under what circumstances are they reviewed? If the application does not provide this traceability, look at the risk assessment relating to this, and ask what additional controls are in place to mitigate this issue.

Also check what is happening in cases of service incidents, such as system failures. Are privileges allocated without due care and attention? Are controls violated? Check the records and logs that are created during such incidents. Check that all privilege activities are monitored and logged, and that someone is responsible for reviewing these logs. Check for each incident reviewed that there is evidence that post-incident restoration of security has occurred, and that a review of actions taken to resolve the incident has taken place.

Access to secure information may also include access to codes for safes and other secure areas. These also need to be recorded and regularly changed. The auditor should also verify that privileges are allocated on a 'need to use' and 'event by event' basis, are immediately removed when they are no longer necessary, and use a different user identity than the one used for normal job functions.

8.3 Information access restriction (ISO/IEC 27001, A.8.3)

> *"Access to information and other associated assets shall be restricted in accordance with the established topic-specific policy on access control."*

Implementation guidance

The owner of the information held in each application, service or solution should specify the access rights and rules to be applied, in accordance with business requirements and the access control policy (see 5.15). These rights and rules should define who will have access and, in the case of information, at what level, e.g. create, read, modify, delete. Without this, there is a high probability that users will be given access to too much information. This creates the risk of a breach of confidentiality and loss of integrity or availability. Over-accessibility can also lead to the risk of fraud in financial applications, and theft of intellectual property.

Be particularly cautious where a shared database is used. Ensure that each role can only access the data meeting the role's requirements, and that applications cannot be used to circumvent the access controls in place.

Dynamic controls can temporarily change privileges, for example to permit third parties to investigate systems faults, or to enable a service desk operative to take control of a user's workstation to help them resolve a problem. Activities taken during periods of changed privilege should be subject to detailed logging, and ideally witnessed by authorised users (e.g. the user of the workstation) to deter misuse. The process should also require explicit permission from an authorised user, and during a shared session, it should be very obvious to the authorised user that their session is subject to dual control or monitoring.

Auditing guidance

Auditors should collect and check relevant evidence to be assured that access to information provided by individual applications is actively managed, matches business requirements and is in line with the access control policy (see also 5.15). For example, different applications might access the same database. Can sensitive information be accessed from one program but not from another? Does this allow the same role different levels of access inappropriately? The access rules in place should also match the handling requirements of the classification scheme (see 5.12).

Users should not generally be presented with lists of information or application functions that they are not allowed to access. Menu options that are not accessible for security reasons should be removed, likewise information in user manuals relating to sensitive functionality. Pay particular attention to little-used parts of applications, such as maintenance utilities. Are these properly controlled?

Auditors also need to check what happens to information that is authorised and accessible. Are there restrictions to what can be extracted as files or in print? How are these extracts then controlled? They should be subject to the same rules regardless of format.

Dynamic privilege management should be clearly defined if in use, along with the circumstances where it is permitted, the process by which it may be achieved and authorised, and the logging or monitoring that must be in place. The auditor should ask to see logs from sessions, and verify that there is a clear authorisation step, and a log of when the privilege started and ended. It should be possible to see which actions on a system were taken by the user, and by other users, while they had changed privilege – people should not be allowing another user to use their online identity.

8.4 Access to source code (ISO/IEC 27001, A.8.4)

> *"Read and write access to source code, development tools and software libraries shall be appropriately managed."*

Implementation guidance

Source code may contain details of the system, other applications and implemented controls. It provides an attacker with a perfect starting point for understanding and performing the unauthorised modification of a system. Serious security problems can result from unauthorised access to, and modification of, source code. Even if source code is well protected, if the development tools (e.g. compilers) are not protected, they can be modified to steal code or alter the function of a program – in short, if the compiler is compromised, all bets are off.

Clear and effective procedures are required to ensure proper maintenance and protection of source code. One method is to use central storage of code. Such repositories should not be held in, and should not be accessible from, production systems, and there should be controls and procedures in place to manage access. Other controls include logging of copies sent, after authorisation, to maintenance staff, logging of updates, strict change control procedures and digital signatures to enable all modifications to be identified. Development tools should be protected to an equivalent level, and only obtained from trusted sources.

Auditing guidance

The auditor needs to check that access to any source code and development tools is protected, ideally by not holding it on production systems and by strictly controlling access to it. Access to such code and the means to modify and re-compile it can effectively bypass the security features within the application. Highly secure applications should include some means of verifying code check sums (or digital signatures). These should be used to identify if unauthorised changes have been made.

The auditor should also check any macros and database report programs. These can be much easier to change and could cause loss of integrity, or make information unavailable. Normal updates of source code should also be properly controlled to prevent installation of the wrong code, and to ensure recording and testing of changes before access to live data is permitted. The auditor should look for documented procedures and records relating to these activities.

ISO/IEC 27002, 8.4 provides additional guidelines that can be applied to secure access to source code and the libraries that this code might use. The auditor should also ask about other standards that are used to secure code.

8.5 Secure authentication (ISO/IEC 27001, A.8.5)

> *"Secure authentication technologies and procedures shall be implemented based on information access restrictions and the topic-specific policy on access control."*

Implementation guidance

Authentication is the moment when the assertion of identity by a user is tested, and the system either accepts their assertion (successful authentication) or rejects it (failed authentication).

While the authentication process should be easy for authorised users to follow, it should not disclose unnecessary information about the operating system, service or application the user is trying to access (e.g. the version number, whether a specific user account exists in the system and error messages revealing database structure). Any information provided might help an unauthorised person trying to obtain access, so the less information given away, the better.

Some systems do not have the facility to control or modify the level of information they present, and the organisation must accept the system as provided. In such cases, other mitigating controls should be used. The organisation should implement what features are available and add further compensating controls where necessary.

The use of passwords alone is not appropriate for many situations. Two-factor authentication (two of 'something you know, something you have and something you are') should be in place and supported by risk assessment. A further third factor can be added to address the most high-risk situations. Be particularly aware that repeated single-factor authentication (e.g. a password followed by another password or PIN) does not provide additional security.

ISO/IEC 27002, 8.5 gives more examples of what a secure authentication procedure should provide. As many aspects of that list as possible should be implemented in the organisation's systems and applications.

Auditing guidance

A risk assessment should have been used to identify the appropriate user identification and authentication mechanisms and procedures. The auditor should obtain evidence relating to the design and implementation of the procedures. Does the user need to carry out a sequence of logon activities? What happens if the wrong information is entered? Is there a delay period, and is a lockout imposed after repeated false attempts? If so, what is the recovery mechanism? Is the password displayed or hidden? Is it obvious how many characters the password should have? Do error messages reveal how the system operates behind the scenes?

ISO/IEC 27002, 8.5 provides a helpful list of properties for secure authentication. Sometimes an application might not provide the necessary level of protection, e.g. if the number of unsuccessful logons is not restricted. In this situation, determine if any other controls have been added, e.g. by physically restricting access to the operating system, enforcing a delay between attempts or steps in a logon sequence, or raising an alert. Verify that the risk assessment has considered each operating system and application in scope, their access methods and authentication procedures, and whether these are considered adequate. Where multiple-factor authentication is identified as necessary, check that its implementation uses different factors, not repeated single factors.

8.6 Capacity management (ISO/IEC 27001, A.8.6)

"The use of resources shall be monitored and adjusted in line with current and expected capacity requirements."

Implementation guidance

With growing requirements for the use of information processing facilities, an organisation may be vulnerable to loss of service because of inadequate resources – both facilities and staff. This risk should be reduced by system tuning and monitoring the use of present resources and, with the support of user planning input, projecting future requirements. Controls detecting problems in capacity can ensure timely corrective action. This is especially important for cloud services and communications networks where changes in load can be very sudden, resulting in poor performance and dissatisfied users.

Capacity should also be managed by managing demand wherever possible to smooth out peaks and troughs. Alternatively, if using externally provided services, it may be possible to arrange for on-the-fly (elastic) increases and decreases in capacity to match demand.

The capacity management process is likely to be cyclical, and evidence of requirements should be obtained and documented in a standard manner that enables reliable repeat capacity calculations to be made. Critical systems, such as network gateways and main database servers, should be prioritised, and a capacity management plan documented for them.

Auditing guidance

Forward planning of basic operational needs is often overlooked, and auditors should check the organisation's ability to handle this. The first question could be, 'What is being monitored?'. This would typically include concurrent sessions, storage usage, network utilisation, and other potential bottlenecks. Other appropriate questions could be, 'What do you use for environment tuning?', 'Where are the logs from the detection controls put in place to identify capacity problems?', 'How do you predict and manage demand?'. Examples of methods for managing demand are given in ISO/IEC 27002, 8.6.

The auditor should enquire how the information received from monitoring is used to identify future capacity requirements. Trending information and extrapolation of future requirements should be being used to plan upgrades. This should include capacity figures, trended as appropriate, reviews and identification of needs and upgrade plans. Look also at staff planning: inadequate human resources at critical times can often compromise security.

8.7 Protection against malware (ISO/IEC 27001, A.8.7)

> *"Protection against malware shall be implemented and supported by appropriate user awareness."*

Implementation guidance

Most (if not all) operating systems are vulnerable to the threat of malware (e.g. viruses, worms, Trojans, etc.). It is very easy for malware to install itself on a vulnerable system, but it can be difficult and costly to get rid of. Prevention can only be achieved to a certain level (new or changed viruses, for example, are often not detected by anti-malware software), but it is still necessary to strictly apply and follow this control. Malicious code in a system can have a devastating impact on the confidentiality, integrity and availability of all files on it, and potentially on any data on linked systems or services that its user has a right to access. Malicious code can spread to any file in the system.

One of the keys to prevention is user awareness (also see 5.10). If staff understand the risks, they will apply the controls and will be wary about what to download, what links to click and which websites to visit. It is vital to also use technical controls that run independently in the background, carrying out checks automatically. A check for malware using a different antivirus tool should also be implemented where appropriate, such as on systems supporting a large number of users.

Another key to success is to ensure that the protection against malware is continually updated. It ensures that protection remains effective, and that relevant information sources are reviewed for news about the latest threats. Updates should be done manually if dynamic updates are not possible for technical reasons.

Auditing guidance

Malware is a problem on almost all operating systems; therefore the auditor should confirm that implemented controls are adequate. The guidance in ISO/IEC 27002, 8.7 can be applied to ensure optimum protection.

A number of options are available when installing software protecting against malicious code. Sometimes it is held on a central server covering all client systems that are logged on. Other systems may require protection software to be installed on each system. Sometimes the installation updates the entire software package, and at other times only libraries are involved, so auditors need to know how to determine the correct versions. Another important item for the auditor to check is that the updating is done whenever it is necessary, for example through an automatic update or some other form of notification of necessary updates. Mobile devices can involve particular problems: how are regular updates assured? Is malware protection available, or necessary, for the device in question?

For systems handling sensitive information, the use of more than one vendor's antivirus software is advisable to improve detection rates.

If checks are not constantly running in the background (very often a good option, but not always possible), procedures should be explicit about regular checking. There should be a clear policy on incoming software, emails and websites. The auditor can also check user awareness of this issue, and look at training records to ensure that users have been provided with information describing correct behaviour.

The actions in the event of virus infection should be covered (also see 5.24). Where automatic cleaning is implemented, ask how the organisation verifies that the cleaning attempt has been successful; in many cases, it may be more advisable to wipe and reinstall a machine, rather than to risk it still being infected. Malware infections should be properly recorded.

Free applications might not give the necessary protection and could cause additional damage. Some malware imitates protection software to extort money from the user. The auditor should confirm whether users know and use the correct methods of interfacing with applications, operating systems, etc. An unexpected request for password information, for example, could be an attempt by an attacker to obtain a password and access vital data.

8.8 Management of technical vulnerabilities (ISO/IEC 27001, A.8.8)

> *"Information about technical vulnerabilities of information systems in use shall be obtained, the organization's exposure to such vulnerabilities shall be evaluated and appropriate measures shall be taken."*

Implementation guidance

Most attacks on organisations' information systems are based on the exploitation of known technical vulnerabilities, and the lead time for these attacks gets shorter and shorter. The organisation should have procedures in place to identify any technical vulnerabilities of its information systems in a timely way, and to identify and implement the appropriate response to such vulnerabilities. Vulnerability scanning tools are widely available and should be used. Manual in-depth penetration testing may also be appropriate for more sensitive or critical systems, as frankly a vulnerability scan will only find surface-level issues (also see 8.29).

The necessary roles and responsibilities for this process should be identified, as well as a timeline for actions, to ensure that appropriate actions are taken within a suitable period.

If a vulnerability has been identified, the organisation should identify the risks related to this particular vulnerability, and the suitable action to be taken, for example whether to install a patch to protect against the vulnerability. If the decision is that a patch should be installed, the organisation should ensure that it is tested before installation, and there should be rollback procedures in place to go back to the previous state if the patch causes unforeseen problems after being installed. ISO/IEC 27002, 8.8 describes the management process for technical vulnerabilities that organisations should consider.

How an organisation evaluates patches before implementation should be consistent with its need for robust information processing systems, and its resources and technical expertise. Small organisations may decide to install

all relevant patches in a timely way (possibly taking account of the experiences other organisations have had), whereas larger organisations might instead carry out extensive in-house testing before the patches are installed. Whatever approach the organisation takes, it should be based on a risk assessment that takes account of the requirements for uptime and resilience to attacks.

Auditing guidance

The auditor should check how the organisation is addressing the issues of published technical vulnerabilities. The organisation should have a structured approach to ensure it finds out about vulnerabilities as quickly as possible, and a process to assess the risks associated with published technical vulnerabilities. To support a decision on what action to take, the organisation should have a clear, documented understanding of its requirements to protect against attacks exploiting technical vulnerabilities, and an appropriate management process for technical vulnerabilities.

The detection process for technical vulnerabilities should include testing procedures, which may include the use of automated vulnerability scanning tools – and in addition to this, manual penetration testing for more sensitive environments. The procedures should be clearly defined and repeatable, and tools should be configured to test all relevant systems, not just a sample. Reports from testing should be available, as well as reports from retests, which verify that vulnerabilities identified in the previous test have been remediated.

Depending on the requirements of the organisation and the resources available, the auditor should check that a sound management process for technical vulnerabilities has been implemented. ISO/IEC 27002, 8.8 describes components of such a management process.

The correct operation of this process should be monitored, and the organisation should have procedures in place to evaluate the effectiveness of its technical vulnerability management. Whatever is done in response to identified technical vulnerabilities needs to link in with the change control process that is in place (see 8.32), and there should always be the option to uninstall a patch and return to the previous state (also see 8.9).

8.9 Configuration management (ISO/IEC 27001, A.8.9)

> *"Configurations, including security configurations, of hardware, software, services and networks shall be established, documented, implemented, monitored and reviewed."*

Implementation guidance

The definition and codification of a system's configuration is a fundamental way to both state what is acceptable (i.e. sufficiently secure) and to detect when unauthorised alterations have taken place. If you know what 'approved' looks like to a good enough level of granularity, then you can identify attacks or unintentional misconfigurations with ease. Obviously, once you have a notion of needing to define 'good', you run the risk of creating a whole industry to create and maintain this definition. This is where externally defined configuration standards, for example those defined by manufacturers or security services, come in – they can do the work of keeping definitions up to date. However, these definitions are, in many cases, one size fits all – they have no notion of local variations in threat, or limitations imposed by the way the organisation's IT works (e.g. a particular third-party tool requiring the use of SSL certificates). The organisation must therefore have a clear view (supported by business case and risk assessment) of which parts of the standard advice will not be adhered to, and why, and what will be done instead.

Some basic good-practice configurations are provided in ISO/IEC 27002, 8.9.

Once the organisation has determined what needs to be placed under formal configuration management, and has identified suitable definitions of 'good enough', it must then ensure that this definition is actively maintained – when a third-party good-practice guide is updated, how will the organisation get enough warning to be able to determine whether to adopt it? The potential issue here is that the organisation has a statement like 'we comply with the CIS Benchmarks for our Windows systems', which is initially correct, but becomes factually inaccurate by the time of audit due to changes in said benchmarks.

Automatic implementation and enforcement of configurations is the lowest-risk approach in terms of compliance, but highest in terms of operational impact; it is a rare organisation indeed that can happily force complete uniformity on all of its systems. The majority may well be able to be dealt with via automation; the edge cases are the area where the majority of the focus of the organisation should be. These are areas where localised risk assessments may be appropriate, and perhaps alternative controls used (remembering that all controls come from requirements of interested parties, or from risk assessment).

Monitoring of conformity with standards is a powerful approach to help detect incidents, whether malicious, accidental or well-intentioned (e.g. a person taking it upon themself to alter configurations to match an upcoming change in a standard). This can then link into incident management processes (see 5.24) and general monitoring activities (see 8.16).

Auditing guidance

The auditor should initially ask what configuration standards are implemented, and to which portions of the organisation's environment. They should then enquire as to where these standards are not currently applied, and how these exceptional cases are addressed. The organisation should be able to produce a risk-based justification for the handling of exceptions – or of configurations that are not adopted at all (e.g. because they are not practical). The mitigations that have been applied to address the risk which is not being addressed (due to the configuration not being adopted) should be clearly defined and appropriate to the situation. For example, a system that must permit insecure certificates might be isolated from the Internet.

Every system to which configuration standards are to be applied should be able to be compared against that configuration standard, and there should be records of this having taken place on a suitably regular basis – ask for justification of the frequency with which configurations are validated. If continuous comparison is taking place, ask to see how alerts are generated, escalated and resolved – are there perhaps so many alerts that they are being ignored?

8.10 Information deletion (ISO/IEC 27001, A.8.10)

> *"Information stored in information systems, devices or in any other storage media shall be deleted when no longer required."*

Implementation guidance

The act of simply getting rid of something once it is no longer needed is a very powerful control – information that no longer exists can no longer be at risk – or pose a risk to the organisation. Some laws and regulations, particularly those pertaining to personal data and payment card data, include deletion as a key requirement.

The primary question to answer is: how will the organisation know which information is no longer needed? This is a question that hides within it a much bigger question: how does the organisation know which information it needs? This second question is potentially intractable, given the amount of information any organisation tends to accrete (e.g. in emails). However, given that the organisation needs to know what it's holding to protect it (a risk assessment cannot be carried out without some appreciation of the information at risk), the asset inventory (see 5.9) can be used as an excellent starting point. How much more granular an asset disposal process needs to be is up to the organisation – it may simply be appropriate to have a retention policy, which states how long certain types (or formats) of information may be retained, with a statement that anything over this age must be securely deleted.

The method of deletion is very much dictated by the budget available, the level of certainty that is needed, the type of storage and the amount of information to be deleted. Physical destruction of media is both very satisfying and reliable; but this assumes that the media in question is no longer required, and, indeed, is known. Where information is stored in the cloud, physical destruction of media is simply impractical. This therefore tends to indicate secure deletion, which comes in a bewildering variety of approaches, from repeatedly overwriting data to the use of cryptography to ensure that information is effectively destroyed (see ISO/IEC 27002, 8.9 for advice on approaches). Custom applications and services should be used as appropriate. Methods should be reviewed regularly and updated as the capability of attackers to retrieve data increases.

Auditing guidance

The organisation should be asked for its information asset register and retention policy; the auditor should then select a random set of assets and ask for records of destruction, checking that these match the policy. Methods for destruction should be consistent with the sensitivity of the information to be destroyed, and again the auditor should ask for a demonstration of the destruction processes (there are likely to be several). There should be evidence that the persons carrying out the destruction know how to do it, and how to verify it has been successful. The auditor should interview a random selection of people handling information that requires special approaches for its destruction – these people should be fully aware that they must use these methods, or the services of specialists, rather than just putting items into the bin, or recycling.

Items awaiting destruction (e.g. old backup tapes) should be treated with the same level of care for their contents as when they were in active use – look for piles of tapes in a corner of the server room, or boxes of old USB sticks on an administrator's desk.

8.11 Data masking (ISO/IEC 27001, A.8.11)

> *"Data masking shall be used in accordance with the organization's topic-specific policy on access control and other related topic-specific policies, and business requirements, taking applicable legislation into consideration."*

Implementation guidance

The guidance in ISO/IEC 27002 8.11 covers not only data masking, but also anonymisation and pseudonymisation, so that's what will be addressed here.

Strictly speaking, the point of masking, anonymisation or pseudonymisation is to ensure that unauthorised individuals cannot see information. The difference between the methods by which this is achieved rests on the level of permanence of the effect achieved. Masking is truly temporary, like holding a paper in front of the information – it's still there, and unaffected, but not visible. Pseudonymisation, which is also called 'tokenisation' in the payments world, replaces a piece of data – perhaps the card number, or a person's name – with a unique string of characters (the token), which is usually randomly generated. A separate 'look-up table' enables an authorised person to look up the token and retrieve the original data. So pseudonymisation renders the data partially unavailable – the look-up table is required. Full anonymisation permanently removes the data; it cannot ever be reinstated. This is surprisingly hard to achieve, as the more anonymous information becomes, the less usable it is. Information can also be 'potentially identifiable' where it does not contain an individual identifier, but can be combined with other information reasonably available to an attacker to derive the identifier. For example, if there are 20 houses in one postcode, and only one bungalow, then any review of health conditions experienced by inhabitants of single-storey buildings that lists results by postcode is potentially identifiable information.

Auditing guidance

In the case of masking or other obfuscation of data, there should be a clear definition of the types of masking, anonymisation, pseudonymisation, etc. in use, and how they should be applied. The roles that are permitted to view the full data (where not anonymised) must be defined, and the process by which the data may be re-combined must be specified. It should be possible to view demonstrations of these processes, and records of where re-combination of data has taken place.

Data that has been re-combined must be treated with suitable care, and deleted as soon as no longer needed (see 8.10), to ensure that it is not retained for convenience. In addition, for pseudonymisation, the look-up table should be in an

environment that is logically and organisationally segregated
from the data containing the tokens.

Ask how the organisation deals with the problem of
potentially identifiable information – is this actively
managed? Are reasonable assumptions made to support the
decisions on how likely this is? It may be worth searching
publicly available information to test whether 'anonymised'
data really is anonymous.

8.12 Data leakage prevention (ISO/IEC 27001, A.8.12)

> *"Data leakage prevention measures shall be applied to
> systems, networks and any other devices that process,
> store or transmit sensitive information."*

Implementation guidance

Data leakage prevention (DLP) is a very challenging topic,
as it consists of preventing data from flowing out of the
organisation without authorisation. This requires the
organisation to know what flows are acceptable, and to
define how these should be secured and authorised.
Unapproved flows are then able to be detected and blocked,
usually by use of one or more DLP tools, which exist in
profusion at the time of writing. The true challenge lies in
first, dealing with false positives, and second, determining
how to approach flows that support a legitimate business
need but are not consistent with the organisation's risk
appetite or requirements (for example, because unencrypted
channels are being used). This then links into the need for a
clear appreciation of the information in scope (see 5.9), and
what needs to be so protected. Users must be aware of what

they can transfer, and where, and how. Flows (e.g. delivery of backup tapes to off-site storage) must be present to support the function of the organisation, supported by risk assessment.

ISO/IEC 27002, 8.12 contains further specific measures that support the management and implementation of DLP.

Auditing guidance

The auditor should first ask what information is intended to be protected by DLP controls; the controls may apply to a specific subset or to the whole of the organisation's information. This should then be traceable to specific systems, networks and other devices that hold this information. The auditor should ask to see how the DLP tools in use are configured to match the priorities of the organisation.

Examples of alerts by DLP tools should be reviewed, and the organisation should be able to show how false positives were identified, and how events were prioritised and addressed; also see 5.25.

Users with access to data within scope of the DLP approach should be interviewed to establish that they are aware of the concept of data leakage, and know what information they should and should not be passing out of the organisation. They should also be able to explain what they would do if they became aware of a potential leakage incident, and what approved methods should be used for sharing certain types of sensitive data, and how authorisation takes place. Ask to see records of the data being transmitted, and matching approval records.

8.13 Information backup (ISO/IEC 27001, A.8.13)

> *"Backup copies of information, software and systems shall be maintained and regularly tested in accordance with the agreed topic-specific policy on backup."*

Implementation guidance

Every organisation is vulnerable to a crashed disk or failed tape, and to other problems that can cause loss or corruption of information or software (e.g. ransomware). Integrity and availability of all important information and software should be maintained by making regular copies to other media. The regularity will depend on the criticality of the data. Some systems can justify real-time backup – writing the copy at the same time as the original; this is particularly useful in cloud environments. If this is not possible, other copying should be used, be it automated or user initiated. A backup cycle should be designed to ensure that all information and software is copied at appropriate intervals, while maintaining at least two copies of each file. Risk assessment should be used to identify the most critical data, which may justify more frequent copying.

Copies should be stored in a safe place. Full copies of data should be kept off-site or at least in a fireproof safe. The protection applied to the backups should be equivalent to that used for the original (e.g. encryption). It is important to regularly test the ability to restore data from backup, and to replace media at end of life. Old media should be adequately disposed of (see 8.10 and 7.10).

Some data will be being kept in a long-term archive for legal or other reasons; this archive may need to be on unalterable media. It is essential to maintain the means to recover data that has been archived. This requires the appropriate computer; media reading device; the software to read the data format, e.g. database manager; and the correct version of the application programs to interpret the data fields. Failure to recover data could leave the organisation in breach of statutory requirements to maintain records. Comprehensive records of tape contents and program/data relationships should be maintained. Alternatively, the data could be transferred to currently supported media as the media it is on is deprecated, if permissible.

Auditing guidance

Backups are a key component in maintaining information integrity and availability. The organisation should have well-defined procedures for dealing with backup. Initially, look at the backup policy and process, which should be consistent with the organisation and its security requirements, and compatible with recovery and business continuity plans.

In a typical physical network environment, this is based on full server backup plus incremental updates defined at a frequency appropriate for the requirements for integrity and availability, for example daily, weekly or monthly cycles. In a cloud environment, snapshots of servers can be taken at suitable intervals, as well as copies of live storage. The auditor should confirm that there are appropriate backup and data restore arrangements in place according to the results of the risk assessment, and check that this includes full coverage of items within the ISMS's scope.

How and where the backup media is labelled and stored is also important. The auditor should check that each item can be uniquely identified, is logged correctly and is held securely. Backup media should be held in separate locations to the systems they back up, and sufficient controls should be in place to give the same level of protection the backed-up information normally has.

The auditor should confirm what the long-term storage requirements of critical data are, and how the organisation validates this. Backup media can deteriorate and may need to be refreshed. Look also in the procedures for backup. What corrective actions are required if the backup fails? What arrangements exist for restoring the data? How often is this exercised? What records are kept? Are backup media tested to ensure they are working properly? Test restores of backup files should not compromise data integrity: check that this is adequately addressed. The auditor should check that requirements for business continuity planning (see also 5.30) are met by the backup arrangements in place in terms of frequency, format and availability.

8.14 Redundancy of information processing facilities (ISO/IEC 27001, A.8.14)

> *"Information processing facilities shall be implemented with redundancy sufficient to meet availability requirements."*

Implementation guidance

Availability is a core aspect of information security, along with other attributes such as confidentiality and integrity. To ensure availability, it is important to identify first what the organisation requires in terms of recovery times following failure, whether it can tolerate any loss of availability and what current systems provide. Where these are not in agreement, additional measures such as redundant equipment and locations should be implemented to increase availability to match need. Risk assessments of the IT and physical architecture of the facility should be carried out to identify and manage any increased risks (for example from copying data between mirrored systems in geographically separate locations via public networks). ISO/IEC 27002, 8.14 contains a list of considerations to assist in the design of adequate redundancy. This obviously also links very strongly to the subjects of ICT readiness for business continuity (see 5.30) and information security during disruption (see 5.29).

Auditing guidance

The auditor should request documents detailing availability requirements, risk assessments, information regarding redundant systems, testing records, records of recovery times following unplanned loss of availability and any compensating controls that may have been introduced to address additional risks incurred by the introduction of redundant equipment. The auditor should confirm that the organisation has appropriate redundancy arrangements that are sufficient to meet availability requirements.

8.15 Logging (ISO/IEC 27001, A.8.15)

> *"Logs that record activities, exceptions, faults and other relevant events shall be produced, stored, protected and analysed."*

Implementation guidance

Logs are a valuable way to detect an incident, and essential to support later investigations. They can establish the events leading up to an incident, and help determine accountability.

However, systems can produce many logs covering a wide range of activities, exceptions and events within the system. Generally, such a quantity of data is difficult to analyse to identify possible misuse; the challenge is sometimes described as 'drinking from the fire hose'. Automated procedures and appropriate tools are therefore vital to distinguish truly significant items from the overall noise of logging information. Data will also need to be cross-correlated between systems to derive actionable intelligence: for example, to trace an IP address to the system using that IP address at the time the event was logged, and then to the user account on that system that owned the process responsible for initiating the event. ISO/IEC 27002, 8.15 contains a list of considerations to be taken into account during log analysis.

The level, type and frequency of monitoring required should be established via risk assessment and focused on the events that matter to the organisation. It is as important as the creation of monitoring logs to assign responsibilities and sufficient time (or service capacity) to review/process these logs; information important for system processing might require more frequent reviews than others. These decisions should be summarised in a clear policy that is consistently applied.

Logs of user (including privileged user) activities should record the items listed in ISO/IEC 27002, 8.15. Logs should be kept for a sufficient period in case they are needed for an investigation. They should also be protected, particularly against unauthorised modification that could eliminate evidence. An ideal approach is to immediately transfer log data to a location that is outside the management of the administrators of the systems generating the log data.

As with other types of incident, system faults can impact information integrity and availability. All faults should be logged to enable suitable corrective action to be taken. In the longer term, logs should be analysed to identify fault trends and identify root causes. The effectiveness of corrective actions should be verified (see also Clause 10.2 of ISO/IEC 27001). Special care should be taken where a fault, or its correction, may have compromised security.

The information in audit logs, administrator logs, fault logs and logs resulting from monitoring activities is only valuable if its integrity can be relied upon. It is therefore essential to give log information sufficient protection. This is especially important when investigating incidents, or when evidence needs to be provided (see 5.28).

The organisation should ensure that it is not possible to:

- edit log files, except with explicit authorisation;
- delete log files, except with explicit authorisation and the creation of a record to document the erasure;

- modify the type of information (e.g. level of event) being recorded in the logs;
- stop logging, except with explicit authorisation;
- redirect log files to an unauthorised destination; and
- overwrite logs (e.g. by using the fact that storage capacity is limited) without generating an alert to an appropriate recipient.

Auditing guidance

The logging policy should specify the administrator and user logs that need to be kept, both for normal operations and fault detection – this should be linked to risk assessment and the requirements of interested parties. In the case of managed services (e.g. cloud services), suppliers will offer a variety of levels of logging; check that the level of logging implemented is appropriate.

Ask for examples of such logs. As a minimum, the log information should identify the event, the person causing the event, any changes made, the date and the time. Transaction codes, terminal ID, network addresses and actions carried out (such as use of privileges or system utilities, changes to the configuration of the system, etc.) may also be recorded. The auditor should review what constitutes a loggable event, and confirm how long this information is required to be held, in what form and under what protection (see also 8.13); a random selection of log entries should be checked to ensure that these requirements are being met in practice. Example events could include changes made by system administrators and failed logins or access attempts. The auditor should check that the logs contain a sufficient level of detail to be useful, and that they are protected from tampering and deletion, especially by privileged users of the systems creating the logs.

The auditor needs to check how often the logs are reviewed – the higher the related security risks are, the more frequently the logs should be reviewed. This check should also confirm whether the review process itself is effective and efficient, and capable of detecting unauthorised changes and deletions. The principle of separation of duties should be employed in reviewing log files; the roles whose activities are being reviewed should not be performing the review.

Tools such as security incident and event monitoring (SIEM) software should be in use to automatically process audit logs, with rules designed to ensure that all relevant incidents and activities are identified, prioritised and highlighted. The auditor should enquire about the use and coverage of such tools, and review how these tools are applied in practice to detect and alert on incidents. The detection configuration should directly trace back to the risk assessment, and events that the organisation needs to be alerted to.

Part of the process for reviewing an alert should be to determine whether security has been compromised. If so, the process should lead to the incident management process (see also 5.24). Check that this process and feed-in is defined in the procedures and authorised by management.

Auditors should ask for a sample of findings that have resulted in corrective action. They should then check that sufficiently authorised personnel have carried out corrective actions, and that these have been effective to mitigate the issue(s) in question.

8.16 Monitoring activities (ISO/IEC 27001, A.8.16)

"Networks, systems and applications shall be monitored for anomalous behaviour and appropriate actions taken to evaluate potential information security incidents."

Implementation guidance

Organisations need to be able to detect the warning signs of an attack, preferably before it turns into a major incident, and in time to respond. Continuous monitoring by an intrusion detection system (IDS), intrusion prevention system (IPS) or security incident and event monitoring system (SIEM) can address this need.

For monitoring to be effective, it is important to know what you are monitoring, how, and why. As mentioned in detail in the previous item (8.15), there is a risk of overload, and a consequent loss of visibility of meaningful intelligence. The organisation should start by identifying what information, systems and services are at risk, and can be monitored. The risk assessment will allow an informed choice of the types of event that need to be detected. These types of event then need to be analysed for each service, system and environment to determine how they can be detected in practical terms. ISO/IEC 27002, 8.16 provides a list of potential data sources and events.

A list of events may be available from the provider of the IDS, etc.; this should always be reviewed in the organisation's operational context to verify that the events relate to those that it cares about. It is common that an event which is highly valuable to detect may be very difficult to configure; customisation is largely inevitable. Correlation of data from multiple sources to detect lateral motion of an attacker who has breached the perimeter is a good customisation use case.

A key challenge in configuring a monitoring system to be useful is determining the threshold at which detections turn into alerts. On top of the 'drinking from a fire hose' problem of having too much data, the events that are being detected may be false alerts (one failed login is not necessarily anything to care about), or may be potential incidents that require significant manual intervention to winnow out the actual incidents (ten failed logins may be a legitimate user with a terrible memory, or an attack). Intelligent use of thresholds is a powerful approach to address this issue (ten failed logins within five seconds, followed by a success from the same source, sounds like an attack). The organisation should use its own risk appetite in conjunction with advice from its monitoring system provider, and from threat intelligence (see 5.7) to fine-tune its detection and alert capabilities, both initially and ongoing.

Auditing guidance

The auditor should ask to see the risk assessment, and trace from it, and from the requirements of interested parties (as relevant), through the risks, to the incidents or events that those risks address. These incidents and events should then be traceable to specific alerts configured in the monitoring system, and then on to the specific sources of information in the logs and data flows from system components. It is important to verify that the monitoring system covers all the systems necessary to detect an incident; some products have blind spots (e.g. they cannot address certain operating systems or infrastructure components) and customisations need to be implemented to ensure that all the relevant environmental components are covered. Some incidents can, however, be detected via indirect means, so a lack of logs from a specific system may not be as large a problem as it appears at first.

Once this 'top-down' review has been carried out, the auditor should also conduct a review of incidents that have been detected via the monitoring system. How have these been detected? Have they been responded to appropriately? What is the approach where a detection requires further verification? Check that there is no evidence of 'event fatigue' where the volume of alerts and detections has resulted in impaired response, potentially impeding detection of incidents.

Finally, the auditor should look at the whole monitoring approach in the round. Does it support the risk appetite of the organisation? Is it reported and followed up appropriately, and does it add value to the protection of the organisation? Given the fast pace of change in this space, the organisation should be continually reviewing its performance in this area, and making changes very actively as opportunities for improvement and issues are detected (also see Clauses 10.1 and 10.2 of ISO/IEC 27001).

A potential area of concern that the organisation should be able to explain is its approach to events that are detected out of hours – the detection team may be there and working, but how do they get in touch with operational staff who can verify an incident, or investigate further? Also see 5.26.

8.17 Clock synchronization (ISO/IEC 27001, A.8.17)

> *"The clocks of information processing systems used by the organization shall be synchronized to approved time sources."*

Implementation guidance

Most output from computers and communications equipment is time and date stamped. This information forms part of the audit trail for transactions moving between computers. It might also be required in investigations or to resolve disputes, and should therefore be entirely consistent between all devices within the scope of the ISMS (see also 5.28). An internal reference time source should be identified, documented and implemented. The use of an external reference time source should be considered, and the decision clearly documented. Radio receivers are available that will provide a computer with a signal from an atomic clock that maintains accuracy to the second. If the organisation chooses to use multiple reference time sources, it should take into account the risk of confusion in retrieval and analysis of incidents if the sources are not consistent, and especially if one drifts relative to another.

Auditing guidance

The organisation needs to establish what its base time will be; if multiple time sources are to be used, then this should be clearly justified and risks identified and mitigated (e.g. inconsistency and drift). For most, the reference time will be local time, but for international organisations some other base may be used, e.g. GMT/UMT. Without proper timing across all systems, audit and monitoring logs can be inaccurate and misleading. Any system audit trails and monitoring investigations rely on accurate system clocks. The auditor needs to verify that access to the configuration of system clocks is restricted to avoid tampering. There should also be some facility to monitor system clocks, and correct them if necessary. Also check what is done to ensure the consistency of wall clocks that may be used for manual logging, e.g. goods received and incident reports. Ensure that these are used, rather than users' watches; are they in the right place? Can system clocks be used instead?

The auditor needs to check how the transition to and from British Summer Time (BST) or other seasonal change is controlled. What arrangements are made to correlate the time for systems that are in different time zones? Are any additional checks carried out when portable/mobile devices log in to the network?

Finally, the auditor should confirm that other systems, such as CCTV, are using the same reference time source as other systems.

8.18 Use of privileged utility programs (ISO/IEC 27001, A.8.18)

"The use of utility programs that can be capable of overriding system and application controls shall be restricted and tightly controlled."

Implementation guidance

System utilities can alter data, expose information and disable systems. These utilities may be necessary to resolve control problems, so their use should be tightly controlled and only take place after sufficient authorisation and justification.

It is essential to identify all system utilities, assess the associated risks and apply measures such as those described in ISO/IEC 27002, 8.18.

Auditing guidance

System utilities may allow access to parts of the system that applications do not normally expose; they may also override or bypass normal security controls. Auditors should collect and check relevant evidence to establish that the organisation has a clear understanding of the utilities installed. Check that they are catalogued and authorised, and that appropriate access control and user restrictions are applied. Check on individual systems that no additional or modified utilities are in existence.

In less well-regulated environments, unauthorised utilities may have been installed that could have consequences not only for the confidentiality of information but also its integrity and availability. The auditor should check that no forgotten utilities remain on systems; in addition to providing unauthorised capabilities, these forgotten utilities may not be being managed, and may hence contain unpatched vulnerabilities (see 8.8), which compounds the problem. Finally, the auditor should check that no one without appropriate authorisation can access or use system utilities. Further controls to secure the use of system utilities are listed in ISO/IEC 27002, 8.18.

8.19 Installation of software on operational systems (ISO/IEC 27001, A.8.19)

> *"Procedures and measures shall be implemented to securely manage software installation on operational systems."*

Implementation guidance

Operational systems are vulnerable to the installation of unauthorised software and unauthorised changes to operational software, with a resulting loss of system and data integrity. Controls are necessary to reduce the risk of system failure, the introduction of any unauthorised software and the possibility of fraud. All software updates should be subjected to change control and authorised and tested before implementation, and all changes should be logged. Backups of old configurations should be retained, and a fallback strategy should be in place for the case of failure of the new system. Staff with installation privileges or development responsibilities should also be trained in the appropriate process for approving and adding new software. Without suitable training, it is entirely possible for competent developers to add unauthorised code to an operational environment without realising it ('it's just a standard library').

New products should be obtained against a business requirement and appropriately authorised. Ensure that valid licences are provided to cover the extent of use intended. For vendor-supplied software, the organisation should ensure that support is available at the level needed, and that this support remains in place for as long as the organisation uses the software.

ISO/IEC 27002, 8.19 provides further guidelines covering the installation and use of software.

Auditing guidance

Auditors should check the controls applied for the implementation of software on operational systems. How is the code held on the system? Is source code included? How are new versions introduced? How are system files and libraries protected? What records are kept of changes? Are changes only made if there are business requirements to do so and after security considerations have been made? Interview developers and maintenance staff to verify that they are aware of the potential dangers of introducing untested code or of allowing unsupported or unauthorised code onto operational systems. Also see 8.18.

The auditor should check implementation. Is this done by suitably competent and authorised staff (also see 8.2)? What protection is applied to source and object code? What testing stages have to be completed before new or modified code is introduced? Is regression testing adequate? Can previous versions of code be reinstalled? Are full backups performed before changes, and are fallback arrangements prepared in case changes have not been successful (also see 8.32)?

8.20 Networks security (ISO/IEC 27001, A.8.20)

> *"Networks and network devices shall be secured, managed and controlled to protect information in systems and applications."*

Implementation guidance

Networks are especially vulnerable to misuse and abuse, as well as unauthorised access or the unintentional failings of technology. They are complex, and it is easy to make mistakes in their configuration, control and protection.

As a result, network integrity can be impaired and availability lost. The confidentiality and integrity of information passing over public networks should also be considered, and appropriate controls implemented to protect the information and the organisation's connected networks and systems, and the information held in these systems.

The only way to reduce these risks is to put in place effective management and security controls, together with sound procedures. Good network security begins with network architecture, and security should be considered throughout the design, implementation, operation, problem management and monitoring. The management of network security has become a significant part of the overall security management activity within an organisation, with specialist knowledge being required for each communications technology. There are also many security controls available to protect the network in different ways, and their use should be properly planned.

- Remote control of network equipment and user workstations for problem solving and software management.
- Network monitors (also known as 'sniffers' or 'taps') to detect attacks and analyse traffic.
- Encryption of transmitted data to retain confidentiality.

- Restricted routing per user or network address.
- Access control techniques to allow only authorised users.

Many of these controls require policies and procedures to be established at the organisational level. All these techniques require comprehensive authorised documentation for network designs, implementation, operation, and changes and monitoring. Constant monitoring of the activities in the network and security status is essential, with appropriate records being kept of faults, problems and corrective actions.

In the case of virtual networks, the risks and measures required are largely comparable with the physical case, but with an additional level of complexity; it is also necessary to consider and manage the risks to the hypervisor or equivalent architectural layer in the infrastructure.

Auditing guidance

Network topology and operating environments, particularly where sensitive traffic is involved, should be properly planned and managed. The auditor needs to confirm that the organisation has done this, and that formal records of these activities are available. Have due consideration and protective mechanisms been employed where networks have access to or use public networks?

For large, complex operations, the use of external experts may be appropriate. If not, look carefully at the qualifications of internal network designers. Have the most exposed aspects of network operations been identified? Has zoning been adequately implemented? What protective measures have been adopted? Security breaches on networks are not always immediately obvious. Data may be intercepted, copied or modified without any obvious trace. Are virtual and physical networks both considered to an equal extent? Are hypervisors and opportunities to bypass the 'canon' architecture in scope and being suitably managed? For example, an 'administrative network' used for managing the network equipment and servers may bypass network zoning. Equally, a backup network connection for use in disaster recovery scenarios may not be as well managed and protected, affording the unpleasant possibility of a security incident happening on the coattails of a service incident (also see 5.29).

The auditor should check what monitoring activities are used to identify and alert on potential breaches, and to confirm that incident reporting procedures cover this (see also 5.24). Network technology is an area of rapid change. The auditor should check how the organisation is monitoring developments in this area and identifying new opportunities and threats.

8.21 Security of network services (ISO/IEC 27001, A.8.21)

> *"Security mechanisms, service levels and service requirements of network services shall be identified, implemented and monitored."*

Implementation guidance

The use of third-party-supplied network services can increase the opportunities for unauthorised access by other parties, leading to losses of confidentiality and integrity.

Availability should also be given special attention, checking on the resilience of the supplier's failover provisions in the event of power, connection or equipment failure The organisation should establish security standards that will be maintained when the supplier is experiencing or recovering from a failure, and which should identify the security features, service levels and controls required by the services being consumed. This is best done via a risk assessment. The organisation should ensure that the identified security features are provided, and documented clearly in the agreement with the service provider (see also 5.20). A list of rules and security features is provided in ISO/IEC 27002, 8.21.

Additional controls may be needed in some circumstances to offset any identified weakness; many service providers offer a standard provision and will not provide additional features on a per-customer basis.

Auditing guidance

Where the organisation is dependent on external suppliers for any networked services, it is essential that the full extent of all available security features, services levels, controls and management requirements is understood. The auditor needs to confirm that the organisation has assessed the risks and the needs for security services, that it has obtained information about the security features from the service provider, and that it has verified that these security features are being provided, and are sufficient and relevant to the identified needs.

The auditor should also check that these security features have been incorporated into operational procedures and security controls, and that there are procedures in place to

review and verify security features regularly. The auditor
needs to confirm that the organisation has covered all
relevant aspects of information security, including
confidentiality, integrity and availability, in its
considerations.

8.22 Segregation of networks (ISO/IEC 27001, A.8.22)

> *"Groups of information services, users and information
> systems shall be segregated in the organization's
> networks."*

Implementation guidance

Networks are vulnerable to unauthorised access attempts,
which can result in breaches of confidentiality and loss of
integrity for the network or its attached systems. The bigger
and more complex and varied the network, the greater the
risk. Security is therefore easier to manage if a large network
is divided into smaller physical or logical domains. Suitable
levels of security can then be provided to manage the
gateways or firewalls between the domains. A perimeter
firewall can also be used to protect the organisation's
networks from unauthorised external access, while still
allowing public inbound access to the organisation's
Internet-facing services (e.g. web servers or API gateways),
and outbound access as required.

Business analysis and architectural modelling should be used
to define the individual network zones required by the
organisation, and risk assessment will determine the level of
security needed to be applied to each zone. Network

connection and routing controls should then be implemented to achieve sufficient segregation of networks.

The zones and their relationships should be carefully documented. The network security plan should be specific about which systems and network devices are in which zone. It is possible for different parts of a single system to be in more than one zone, for example by department or business unit. Provided that the security system can logically segregate them, this may be acceptable.

Auditing guidance

The auditor should:

- enumerate what network zones the organisation has put in place;
- establish that they are appropriate for the organisation's security requirements;
- determine how they are defined and incorporated into network operations; and
- verify how the connection from one network zone to another is controlled.

Considering that security zones can impose restrictions in operational performance, the auditor should also investigate whether performance-related modifications, especially to the implemented environment, have led to any security compromises, and whether performance is monitored (to detect and hopefully prevent availability incidents).

Where wireless networks are in use, the auditor should check that a risk assessment has been carried out to determine whether direct connection between the wireless network and

the main network is appropriate, and whether this assessment
has identified and implemented all relevant controls to
manage the risks. The organisation should implement known
good security standards when configuring any wireless
network, especially if it is directly connected to the main
network.

8.23 Web filtering (ISO/IEC 27001, A.8.23)

> *"Access to external websites shall be managed to reduce
> exposure to malicious content."*

Implementation guidance

A significant way to compromise a target system is to trick
the user into either visiting a site that automatically installs
malware or downloading and installing malicious software
themselves (e.g. fake 'AV' software). To combat this, user
education is vital, but no one is infallible. Additional tools
are valuable to further reduce the chances of users visiting
hostile websites. Tools to block access to known bad sites
can be installed on the local machine or implemented in the
network infrastructure (or both – but this may be confusing
if different lists are used).

The challenge is to identify what 'known bad' is – and
services are available, usually as part of the tools, that
dynamically categorise sites based on both security and
acceptable use. The organisation should decide which
categories should be blocked, based on its risk assessment,
and what to do with sites that have not yet been categorised.
This will extend beyond information security to other aspects

of permitted use. See ISO/IEC 27002, 8.23 for a list of categories to consider blocking.

A policy should be defined regarding what is permitted, supported by the blocking tool, and staff should be trained in acceptable use (also see 5.10). Different policies may be implemented in areas with higher requirements for protection; it is a balance between protection and usability.

Once a site has been categorised and blocked, the organisation can choose to report on attempts to reach that site – for example, if a site is known to be providing command-and-control functions, this detection may indicate that the system attempting to reach it has been infected by malware. This type of event can be used in the monitoring system (see 8.16).

Auditing guidance

The organisation should be asked to provide its policies regarding web filtering, and to show how these address risks such as malware, misuse and phishing. The process for ensuring that the filtering is consistent with policy, and applied to all in-scope systems, should be clearly understood and implemented, and appropriate exceptions applied, with a process for reporting of false positives and requesting further exceptions. Appropriate and consistent decisions should be made regarding exception handling, and authorised. The auditor should verify that it is not possible for staff to manually bypass or disable the web filter, and that attempts, where possible, generate alerts that are fed into the monitoring system (see 8.16).

Ongoing reviews should take place to ensure that web filtering is continuing to be effective, and is not blocking legitimate business activity. The auditor should interview staff to verify that they understand what filtering is in place, and why.

Attempts to connect to known command-and-control servers, or similar destinations, should link into the incident handling process, possibly via the monitoring process.

8.24 Use of cryptography (ISO/IEC 27001, A.8.24)

> *"Rules for the effective use of cryptography, including cryptographic key management, shall be defined and implemented."*

Implementation guidance

The effective use of cryptographic techniques is only possible if basic principles are identified, agreed and applied. For example, the algorithms used should be suitable for the business processes and services they are supporting, the key length should be appropriate for the security requirements of the information that will be protected, and the solutions implemented should be consistent throughout the part of the organisation where cryptographic controls are applied. The organisation should avoid designing its own cryptographic algorithms unless that is its only activity; it is easy to create a weak but apparently strong algorithm. Relevant legislation on what algorithms are permitted should also be taken into account (see 5.31).

To achieve this, a risk assessment should be used to determine the requirements for confidentiality, integrity, authenticity and non-repudiation, and the most suitable cryptographic solutions and a policy on the use of cryptographic controls should be communicated to all users of such controls before any application. This policy should take into account the relevant key-management activities and legal issues involved in the use of cryptographic techniques.

The key-management system used should provide protection of the cryptographic keys according to their use, and management methods that support the handling and use of keys as required by the business processes for which the cryptographic controls will be used.

The requirements for key management will be different depending on which cryptographic technique, secret or public key technique will be used, and what type of key, public or private, is considered. The protection of secret and private cryptographic keys is different from the protection necessary for public keys. When defining a key-management system, these protection requirements should be analysed with the help of a risk assessment, and appropriate protection should be in place before the first production keys are generated and used.

A set of standards and procedures for the key-management activities as described in ISO/IEC 27002, 8.24 should be agreed and implemented before using cryptographic controls. The lifetime of cryptographic keys should be defined for each application in relation to the risks of possible damage if they are compromised, and the deactivation of keys should take place immediately when they reach this age.

The organisation should also consider its needs for keeping copies of keys or parts of keys used for cryptographic controls, either for its own use or to satisfy legal requirements. It might be necessary to agree public-key-management processes with certification authorities, and the organisation might also want to consider the use of other services such as key generation, distribution and revocation, directory services or timestamping, which may be offered by third-party organisations.

Finally, the organisation may choose to retain copies of the employees' keys to avoid any misuse, such as the unauthorised distribution of the organisation's information, or a disgruntled employee first encrypting information and then destroying their private key.

Auditing guidance

An important aspect of the secure and effective use of cryptographic controls is to make sure that the requirements have been identified, and that the correct decisions have been made about what cryptographic controls to use. A policy should be in place supporting the day-to-day use of these controls. This policy should cover the key-management approach applied (see ISO/IEC 27002, 8.24 and below for examples), the roles and responsibilities related to the use of cryptographic controls, and the information and circumstances for which cryptographic controls should be applied.

If the organisation applies cryptographic controls, auditors should check that a policy on the use of cryptographic controls has been developed and communicated, and is known and followed by employees. The auditor should verify that the decisions shaping the policy are traceable to a risk assessment, and that the controls implemented are commensurate with the policy.

The strength of cryptographic controls does vary, and is related to the algorithms employed and the key sizes and parameters used. A factor to be taken into account is the environment and application in which cryptographic controls are applied. Some application environments might require the use of stronger cryptographic controls. Equally, as the implementation of stronger cryptographic controls may have an impact on performance, it is necessary to also consider this aspect when reviewing policy decisions.

The auditor will need to have at least a general knowledge of cryptographic techniques and mechanisms, key management, and their implementation to assess whether what the organisation is using is adequate and appropriate. The use of specialist technical expertise may also be necessary to support the auditor in this area. A useful rule of thumb is that all algorithms must be industry-standard and not known to have been broken. Internally devised cryptographic algorithms are almost without exception fatally flawed.

Key management is a prerequisite for the secure use of cryptographic controls, and no cryptography should be used without a secure key-management system in place. The whole lifecycle should be considered (see ISO/IEC 27002, 8.24 for a list of stages in the life of a key).

Auditors should check that the organisation has implemented adequate controls to protect:

- secret and private keys against replacement, disclosure, modification and destruction; and
- public keys and public-key certificates against replacement, unauthorised modification and destruction.

If the organisation is using a certification authority (public or internal) for the management of its public keys, it should be able to show how it has verified that the authority is trustworthy and suitably managed.

The protection of cryptographic keys should encompass both logical and physical protection. Auditors should review the physical and logical access controls that are being applied to protect cryptographic keys. Where keys are managed by, or in combination with, third parties, the organisation should have agreements in place that cover key protection.

In addition, auditors should check that other relevant key-management procedures, for example as described in ISO/IEC 27002, 8.24, are in place. If a key-escrow arrangement (or other provision for emergency recovery) is in place to protect cryptographic keys, the auditor should ensure that the appropriate employees are aware of it, and that there are no possibilities to circumvent key escrow or to inject an unauthorised key.

8.25 Secure development life cycle (ISO/IEC 27001, A.8.25)

> *"Rules for the secure development of software and systems shall be established and applied."*

Implementation guidance

Where software, services, networks or whole environments are being developed, the organisation should consider the information security of these environments to prevent the deliberate or accidental inclusion of inappropriate functionality (or of vulnerabilities) that could be used later in the live system to compromise it, and to protect the intellectual property of the organisation, both embodied in the materials being developed, and in the tools and systems being used in the development. The compromise of a development environment itself may impact the confidentiality, integrity and availability not only of that environment, but also of the production environment in the future.

A policy should be implemented to ensure that development is carried out to standards that are suitable to the organisation's risk profile. ISO/IEC 27002, 8.25 contains a checklist of what should be in the secure development policy; the list summarises measures that are covered in more detail in other controls in the Standard.

Auditing guidance

The auditor should look for evidence of secure coding or other relevant standards in the development environment, and ask to see documentation supporting a consistent and suitable approach to the identification and resolution of vulnerabilities. Developers should have training and competence in secure practices, which can usually be ascertained via interview. The auditor should confirm that information security standards are not only used, but also that adherence to them is mandated and checked. Third parties (e.g. external organisations or contractors) should be required to meet the same rules as internal employees. The auditor should confirm that the organisation formally verifies compliance on a routine basis, and follows up and resolves non-compliances by any party. The auditor could look for records of secure coding audits, both against standards and to find actual vulnerabilities. Check that these were not performed by the staff who wrote the code in question, and that they were performed by appropriately competent individuals. Automated code audit tools will find many issues, but should not be relied on where coding standards are inconsistently applied. Finally, there should be a regular process to review coding standards for effectiveness – have public standards evolved? What updates need to be made? Also see Clause 10.1 of ISO/IEC 27001.

8.26 Application security requirements (ISO/IEC 27001, A.8.26)

> *"Information security requirements shall be identified, specified and approved when developing or acquiring applications."*

Implementation guidance

Applications are a core part of the way any organisation operates, from email, websites and industrial control systems, through to applications that the organisation itself may produce for use by its customers.

ISO/IEC 27002, 8.26 provides a range of requirements that should be applied to the selection and development of applications (also see 8.25). For example, the application of cryptographic controls (see also 8.24) can achieve protection in several ways:

- encryption can be applied to ensure the confidentiality of information such as billing details, customer information and personal information;

- digital signatures can be applied to ensure the integrity of electronic transactions and to authenticate the partners involved in transactions; and
- encryption and digital signatures can be used to achieve non-repudiation that helps resolve disputes regarding the occurrence or non-occurrence of events.

When using cryptographic controls, care should be taken to ensure that an appropriate policy and key-management system is in place, and that these controls conform to any legal requirements (see also 5.31) that might be applicable.

Any organisation providing or using transactional services over public networks (e.g. payments) should have a policy in place that describes who is allowed to manage these services, what each of these employees is authorised to do, and what controls are in place to protect and monitor such activities. The organisation should use a risk assessment to identify the level of protection required for information involved in transactions. ISO/IEC 27002, 8.26 describes items that should be considered for application services transactions, depending on the concerns of the organisation.

Auditing guidance

Auditors should enquire about the current and future activities within the organisation. All activities related to the organisation's use or provision of applications should be reviewed for any security-related aspects. This includes a check of the following.

- Is an authorisation process in place? Can only those employees within the organisation who are authorised carry out activities?

- Is there suitable segregation of duties? Are activities that, in combination, can be used to commit fraud or otherwise compromise legitimate communications segregated or supervised?

- Are appropriate cryptographic controls in place (see also 8.24) to ensure the authenticity, integrity and confidentiality of information processed in relation to non-repudiation of actions and events, and is a policy in place to regulate the application of such controls?

- Are appropriate network security controls in place to protect the organisation's network and the host used for public application services from attacks that can result from the interconnection with other networks (see also 8.20)?

- Are procedures applied to achieve appropriate verification of actions, payments, etc.?

- Have actions been taken to arrange sufficient insurance?

- Is sufficient protection given to guard against risks from other security problems that might relate to information involved in the provision of public application services?

The auditor should check how the organisation has addressed the issue of identifying and implementing the appropriate level of protection for transactions. Has the organisation carried out a risk assessment? The assessment should take account of at least the following issues, in addition to the list above.

- The secure use of, and communication with, authorities managing certificates for digital signatures.

- Verification of who is involved in the transactions and of user credentials.
- The use of secure communication protocols.
- The secure storage of the information involved in transactions.
- Compliance with applicable legislation and/or regulations, depending on the jurisdiction(s) that might be involved in the transaction.

8.27 Secure system architecture and engineering principles (ISO/IEC 27001, A.8.27)

> *"Principles for engineering secure systems shall be established, documented, maintained and applied to any information system development activities."*

Implementation guidance

The discipline of security engineering is a relatively mature one, and provides a whole raft of principles and tools that can usefully be applied to the implementation of any information system (e.g. zero trust, which is particularly popular at the time of writing). It enables security to be designed in from the beginning, which makes information security cheaper and more likely to be viable (retro-fitted security is never the best plan). It also integrates information security into standard development and implementation activities, which is critically important. Principles and processes to consider are given in ISO/IEC 27002, 8.27.

Security engineering approaches must be tailored to the environment and to the organisational culture. They should also be regularly reviewed to ensure that they are up to date and achieving the desired goal of making information security intrinsic to information system engineering activities.

Auditing guidance

The auditor should check what security engineering principles the organisation uses when designing and creating an information system, where it obtained them and how it keeps them up to date. What assessment has been done to verify that the principles are suitable, and are they implemented as described? How are they promulgated to contractors and external parties? There is likely to be a role with responsibility for security engineering. Interview staff assigned this role to ascertain their level of competence in the field.

8.28 Secure coding (ISO/IEC 27001, A.8.28)

> *"Secure coding principles shall be applied to software development."*

Implementation guidance

This control supports the high-level control 8.25, Secure development life cycle, but focuses on the detail of secure coding standards.

Consistency in coding has multiple advantages: it avoids having to reinvent coding practices, it ensures portability, and it supports ongoing maintenance and support of existing code, if the coding standard is well chosen. Over and above these general benefits, secure coding standards aim to ensure that code is not only consistently structured but also consistently secure. The basic security pitfalls in a given coding language (e.g. memory management) can be addressed by adopting a coding standard designed for that language. High-level coding standards will specify generic principles (e.g. 'do not hard-code passwords'), while granular coding standards provide detailed instructions, including lists of functions that should (or should not) be implemented for a higher level of trust in the code being created. High-security code will have far more restrictions in its design than code designed for a purpose that does not require high levels of security.

The adoption of a secure coding standard becomes even more powerful when combined with automatically generated code (e.g. AI-written); the use of these standards makes it more likely that the code will be comprehensible to humans, can be verified and therefore trusted. Automated review tools that test code for compliance should also be implemented.

Secure coding processes should be used to support coding standards; examples of these are listed in ISO/IEC 27002, 8.28.

Auditing guidance

The auditor should begin by establishing what coding is currently taking place within the scope of the ISMS (carried out either by members of the organisation or by third parties). For each area/context where coding is happening, the organisation should be able to demonstrate how it has determined the risks associated with the activities, and how it has selected the secure coding standard(s) and processes to be implemented. The standards and processes should be clearly documented and available to all parties who need to use them, or to verify that they are in use. The persons carrying out coding should be interviewed to verify that they know of the existence of these documents, know which one(s) apply to their work, and can retrieve the relevant documentation explaining what they need to do to comply with the Standard and supporting processes. They should also know what to do if they encounter a situation where the Standard cannot be applied (e.g. because of constraints from legacy applications). Finally, the persons responsible for internal audit should be asked to show how they verify that coding has been carried out to the practices and standards applicable; it is advisable to carry out this last verification step, given the highly technical nature of this control.

8.29 Security testing in development and acceptance (ISO/IEC 27001, A.8.29)

> *"Security testing processes shall be defined and implemented in the development life cycle."*

Implementation guidance

Security testing should, of course, be carried out prior to deployment. It is important that acceptance criteria are established, and checked to ensure that vulnerabilities are identified and eliminated. This control is also applicable where new subsystems and devices are being introduced, and where changes are being made to existing systems.

To ensure that vulnerabilities and inappropriate functionality (e.g. Trojans) are removed from code as soon as possible, and not left until the end of the development process, systems should also be tested on an ongoing basis.

Testing may include vulnerability scanning, penetration testing, automated source code testing, formal methods analysis and manual code review, as relevant and appropriate. ISO/IEC 27002, 8.29 suggests some types of testing, including security testing, that may be carried out. All levels of acceptance testing should be documented and signed off by an appropriate role.

For major new developments, the operations function of the organisation and other relevant stakeholders should be consulted at all stages in the development process to ensure the practicality of the proposed security design. Involving users is important, as they need to operate the system securely as part of their work. Appropriate tests should be carried out to confirm that all acceptance criteria are fully satisfied.

Auditing guidance

The auditor should check whether system testing records exist of testing of systems throughout their development. The frequency of testing should have been determined by a risk assessment during the design phase, and should be revisited if there are issues found resulting from insufficient testing. Tests may be performed by the team carrying out development, in the early stages of work.

The auditor should look for clear acceptance criteria for security that need to be fulfilled before implementing new or upgraded systems. New systems or processes need to be thoroughly tested before operational use. What plans are there? Have they been reviewed for adequacy? How have the results been recorded?

Adequate testing usually means more than just testing new functionality. Has sufficient consideration been given to regression tests? Has the system response to defective data or false user input been covered? Are access controls fully secure? What about other security controls?

Training may need to accompany system acceptance to ensure that the system is used securely. Has this been catered for? Who has determined its adequacy? Have all necessary personnel been involved, both in the preparation and receipt of training? Who authorises final acceptance before operational use? Check this is defined and recorded.

Has user testing investigated whether, and how, users may seek to bypass security measures?

8.30 Outsourced development (ISO/IEC 27001, A.8.30)

> *"The organization shall direct, monitor and review the activities related to outsourced system development."*

Implementation guidance

Outsourcing software development carries risks relating to the lack of direct visibility of the development process to the organisation. These risks include inappropriately low-quality products, as well as unwanted software, such as covert channels or Trojan code being integrated into the product. Clear contractual agreements should be used to protect against these risks to ensure the timely delivery, sufficient quality and reliable functioning of the software, and to identify the intellectual property rights of the work carried out (see 5.20). In addition, checks should be devised (as appropriate and possible) to verify that the development carried out is as specified. This may include testing of deliverables, auditing of the outsourcer's environment and obtaining evidence to support the outsourcer's compliance with the organisation's requirements (e.g. outputs of testing, lists of flaws identified and addressed, and certificates to confirm training of developers in secure development).

ISO/IEC 27002, 8.30 lists a range of considerations to be taken into account when contracting with third parties, both directly and indirectly.

Auditing guidance

The auditor should confirm that the risks of outsourcing software development are being considered and assessed on an ongoing basis (also see 8.28, 8.29, 5.19 and 5.20). Ideally, the organisation should inspect and review the development environment and processes involved. The auditor can then check the results. The organisation's risk assessment should also consider indirect risks through the supply chain – what if a supplier handling the organisation's information uses insecure software?

The auditor should also check that risks associated with such developments are periodically reviewed, and that changes to security requirements, controls and responsibilities of both parties relating to such developments and any changed or new risks are covered by contract (also see 5.20). The auditor should check that contracts cover:

- conditions to measure the timeliness and quality of developed software and requirements for the quality of code;
- access rights in case an audit is necessary to ensure quality of work done;
- regulations and agreements defining intellectual property rights and ownership of developed software;
- sufficient testing of the functionality of developed code, including checks for viruses, covert channels and Trojan code; and
- training requirements for developers.

8.31 Separation of development, test and production environments (ISO/IEC 27001, A.8.31)

"Development, testing and production environments shall be separated and secured."

Implementation guidance

Production environments require appropriate levels of integrity and reliability. Risks are particularly high where new operational software, communications equipment or services are being developed or tested. Errors and omissions can lead to unauthorised access, introduction of malicious code and other security problems.

Where an organisation is carrying out development activities, it must not only have policies to support suitable processes (see 8.25) but also a suitably secure development environment (see 8.22). As with any environment, the risks may come from not only the technology being inappropriately selected and configured but also from the actions of people with access to the environment. Multiple development environments may be required to separate more sensitive work from less sensitive work. The organisation may indeed choose to have an entirely segregated environment for every development activity it carries out, to reduce the risks of cross-contamination. In this case, there may be a common template for each category of environment to simplify the set-up process. Where the work involves integrating multiple systems, the scope of the work, and hence the risks, are higher. Considerations that should be taken into account are listed in ISO/IEC 27002, 8.31.

It is highly inappropriate to put sensitive data into a test environment unless this is as secure as the live environment, especially since, as you are testing something, it is far more likely to have security flaws. Hence any security controls that form part of the application or other entity being tested should not be used to protect the test environment.

Measures such as strong access control should be applied to separate development, test and operational facilities. The easiest way is to use entirely separate systems, or at least separate domains that are completely segregated from each other. Where such a separation cannot be completely achieved, separate logon procedures supporting access control and good monitoring (see 8.16) can be implemented to achieve similar effects. Tested and verified solutions should be passed to the change control procedure for acceptance into production (see 8.32).

Auditing guidance

It is important that appropriate separation between test, development and live environments is achieved. The auditor should check how such separation is implemented, and what authorisation processes ensure that development and untested application software is not used on operational systems. The auditor should also review the risk assessment to verify that this issue has been given appropriate consideration, and that the controls in place are adequate to protect against the identified risks.

If operational applications software and information are held on the same system as those under development and in test, then the auditor should check that they are at least held in separate domains, and that strong access controls are in place to ensure that no merging of development, test and operational facilities takes place.

The auditor should ask to see documentation describing the secure development environments in use. They should also ask about principles used to determine appropriate controls to apply to each environment, and what (if any) connections may exist between environments. A revealing subject to explore is that of administration. Does the design of the development environment take into account the risks posed by any common administration back-end, such as a private administrative network or authentication service?

The auditor can review designs and implementations against the items in ISO/IEC 27002, 8.31. Interview users of the environment(s) to see if they are aware of the measures that should be in place, and any behaviours that they are expected to exhibit. What background checks are carried out on staff working in the secure development environment(s)?

If sensitive data exists in the test environment, the security measures there should be at least as good as those in the live environment, neglecting any measures that are part of the system/software being tested. Additional security controls may be needed for the test environment to take account of this.

8.32 Change management (ISO/IEC 27001, A.8.32)

> *"Changes to information processing facilities and information systems shall be subject to change management procedures."*

Implementation guidance

Uncontrolled changes to the organisation, its processes, applications, information processing facilities and systems can cause major interruptions to business processes. Changes that might cause problems include the installation of new software, changes to a business process or operational environment, acquisition of a new business, or the introduction of new connections between information processing facilities and systems.

Formal control and coordination of all changes should be implemented, together with business and technical authorisation, for each change at all stages of development – requirements capture, design, code, test, and transition to operational status. Changes should be planned and prepared with appropriate testing and review, and the application and operational change control procedures should be integrated and linked as much as possible. The organisation should sign off final testing before operational implementation. ISO/IEC

27002, 8.32 provides further information about the change control procedures that should be applied.

To avoid interruption to business activities, any changes to operational systems should only take place after the necessary testing has been carried out and formal approval has been given. The procedures for such an approval should take into account the potential impacts, including on security controls in place.

The change procedure should also allow for failback arrangements, for stopping changes, if necessary, and defining what action is needed to recover from unsuccessful changes. All changes that are made need to be fully documented, for example in an audit log that contains all relevant information.

Auditing guidance

The auditor should check that management responsibility and formal procedures are in place to control changes to operational information processing facilities, including systems, infrastructure, networks, applications, code, etc. All such changes should be monitored, and logs should exist describing exactly which changes have been made. The auditor should confirm that no changes can take place without first assessing the possible information security impact that such changes may have, and obtaining appropriate approval for the proposed change.

The auditor should check that procedures are in place describing how to monitor a change for issues while it is being carried out and how to react if something goes wrong. The auditor should confirm that no change can take place without appropriate fallback procedures allowing a return to the original state. The auditor should also confirm that the procedures cover informing all relevant personnel when a change has taken place. A good indication can be obtained by not only looking at the change control procedures but also at records of previous changes to check that these records contain all necessary information and support the evidence that the change control procedures have been complied with. The auditor should confirm that there is a complete history of all changes made and that these records are retained for as long as is required, as well as a copy of the original (as applicable).

If operational changes also yield changes to the applications, the changes should be integrated (see also 8.32). Changes should also result in changes to associated process or specification documents, and tests should include tests for vulnerabilities that may have been introduced.

Changes will often be grouped together and incorporated into a release rather than introduced separately. In this situation, the auditor should look to see that release records correctly identify each of the changes made, that proper configuration control is applied during all changes and that correct records of the implemented release are in place.

It is likely that an emergency change procedure is also employed to correct operational system failure situations. Check that this also meets all the above criteria.

8.33 Test information (ISO/IEC 27001, A.8.33)

> *"Test information shall be appropriately selected, protected and managed."*

Implementation guidance

Test data should normally be fictitious, but there are
occasions when operational, tokenised or anonymised
operational data needs to be used (also see 8.11). The
organisation is vulnerable to breaches of confidentiality
when such data is used and it should avoid such use as far as
possible, and control and protect the data to at least the same
extent as operational data in an operational system if its use
in tests cannot be avoided. The use of operational data in
testing should be recognised in risk assessments, and the
higher security requirements noted in test plans. Each
instance of use should be authorised, and the same level of
access controls as applied for the operational system should
be in place. When the tests have been finalised and the data
are no longer needed, they should be erased immediately and
securely from the test system(s). ISO/IEC 27002, 8.33
suggests further measures to manage the risk of using live
data in test environments.

Auditing guidance

The auditor should confirm, by means of appropriate
evidence, that any live (production) data used for testing is
properly controlled. Tests should be reproducible and the
data used should be distinct and available for any retesting.
Use of live data for testing should be discouraged and, if
used, it should be modified to remove any personal or
otherwise sensitive information. This is not always possible
– or not possible to completely assure – so check how this is
handled and how any results of the testing – data files, logs,
debug files, caches and recorded results – are protected.
There may also be legal requirements relating to the use of

personal data for testing (e.g. a requirement to anonymise the data, or get explicit permission from the people to whom it pertains) that should be fully explored and satisfied.

The use of live data for testing should be properly authorised on each occasion – check that this is done. Check also that there is a method to completely remove any data put into live databases during testing, before accepting solutions into production, and verify the access control in place (also see 8.31). Access to test application systems should be as tightly controlled as access to operational systems. All actions carried out during the tests should have been logged – check that these logs exist, and use them as evidence of how test data is handled and protected.

8.34 Protection of information systems during audit testing (ISO/IEC 27001, A.8.34)

> *"Audit tests and other assurance activities involving assessment of operational systems shall be planned and agreed between the tester and appropriate management."*

Implementation guidance

Before any audit of information systems takes place, the audit requirements should be assessed and, if an audit is required, it should be carefully planned and a schedule agreed. Audit activity on operational systems may require the use of special programs that access data files used by the system or its applications. Such use should be planned to avoid causing problems and disruption in operational systems. Audit plans should be documented and authorised.

ISO/IEC 27002, 8.34 gives further guidance on conducting information systems audits.

Auditing guidance

Information system audit controls and tools used in the audit should not compromise either the information or operations being checked. Where audits are planned, check that the requirements have been identified and appropriate authorisation has been obtained from operational management. No information should be changed for the purpose of these activities, and access to information should be logged as for any other operation.

The auditor should also check that the interruptions to business activities are minimised. Make sure the audit results are kept and that use of any tools is properly recorded. A check can also be made that any tools are themselves formally validated before use. This includes checking that the person carrying out the audit is independent of the activities being audited.

FURTHER READING

IT Governance Publishing (ITGP) is the world's leading publisher for governance and compliance. Our industry-leading pocket guides, books and training resources are written by real-world practitioners and thought leaders. They are used globally by audiences of all levels, from students to C-suite executives.

Our high-quality publications cover all IT governance, risk and compliance frameworks and are available in a range of formats. This ensures our customers can access the information they need in the way they need it.

Other publications you may find useful include:

- *Artificial intelligence – Ethical, social, and security impacts for the present and the future* by Dr Julie E. Mehan, *www.itgovernance.co.uk/shop/product/artificial-intelligence-ethical-social-and-security-impacts-for-the-present-and-the-future*
- *ISO 22301:2019 and business continuity management – Understand how to plan, implement and enhance a business continuity management system (BCMS)* by Alan Calder, *www.itgovernance.co.uk/shop/product/iso-223012019-and-business-continuity-management-understand-how-to-plan-implement-and-enhance-a-business-continuity-management-system-bcms*

- *IT Governance – An international guide to data security and ISO 27001/ISO 27002, Eighth edition* by Alan Calder and Steve Watkins, *www.itgovernance.co.uk/shop/product/it-governance-an-international-guide-to-data-security-and-iso-27001iso-27002-eighth-edition*

For more information on ITGP and branded publishing services, and to view our full list of publications, visit *www.itgovernancepublishing.co.uk*.

To receive regular updates from ITGP, including information on new publications in your area(s) of interest, sign up for our newsletter at *www.itgovernancepublishing.co.uk/topic/newsletter*.

Branded publishing

Through our branded publishing service, you can customise ITGP publications with your company's branding.

Find out more at

www.itgovernancepublishing.co.uk/topic/branded-publishing-services.

Related services

ITGP is part of GRC International Group, which offers a comprehensive range of complementary products and services to help organisations meet their objectives.

For a full range of resources on information security visit *www.itgovernance.co.uk/shop/category/information-security*.

Training services

The IT Governance training programme is built on our extensive practical experience designing and implementing management systems based on ISO standards, best practice and regulations.

Our courses help attendees develop practical skills and comply with contractual and regulatory requirements. They also support career development via recognised qualifications.

Learn more about our training courses and view the full course catalogue at *www.itgovernance.co.uk/training*.

Professional services and consultancy

We are a leading global consultancy of IT governance, risk management and compliance solutions. We advise businesses around the world on their most critical issues and present cost-saving and risk-reducing solutions based on international best practice and frameworks.

We offer a wide range of delivery methods to suit all budgets, timescales and preferred project approaches.

Find out how our consultancy services can help your organisation at *www.itgovernance.co.uk/consulting*.

Industry news

Want to stay up to date with the latest developments and resources in the IT governance and compliance market? Subscribe to our Weekly Round-up newsletter and we will send you mobile-friendly emails with fresh news and features about your preferred areas of interest, as well as unmissable offers and free resources to help you successfully start your projects. *www.itgovernance.co.uk/weekly-round-up*.

EU for product safety is Stephen Evans, The Mill Enterprise Hub, Stagreenan, Drogheda, Co. Louth, A92 CD3D, Ireland. (servicecentre@itgovernance.eu)

www.ingramcontent.com/pod-product-compliance
Lightning Source LLC
Chambersburg PA
CBHW042310210326
41598CB00041B/7334